The Economic Utilisation of Food Co-Products

RSC Green Chemistry

Series Editors:
James H Clark, *Department of Chemistry, University of York, UK*
George A Kraus, *Department of Chemistry, Iowa State University, Ames, Iowa, USA*
Andrzej Stankiewicz, *Delft University of Technology, The Netherlands*
Peter Siedl, *Federal University of Rio de Janeiro, Brazil*
Yuan Kou, *Peking University, People's Republic of China*

Titles in the Series:

How to obtain future titles on publication:
A standing order plan is available for this series. A standing order will bring delivery of each new volume immediately on publication.

For further information please contact:
Book Sales Department, Royal Society of Chemistry, Thomas Graham House, Science Park, Milton Road, Cambridge, CB4 0WF, UK
Telephone: +44 (0)1223 420066, Fax: +44 (0)1223 420247
Email: booksales@rsc.org
Visit our website at www.rsc.org/books

The Economic Utilisation of Food Co-Products

Edited by

Abbas Kazmi
University of York
Email: sk_abbaskazmi@yahoo.com

and

Peter Shuttleworth
CSIC, Spain
Email: peter@ictp.csic.es

RSC Publishing

RSC Green Chemistry No. 24

ISBN: 978-1-84973-615-2
ISSN: 1757-7039

A catalogue record for this book is available from the British Library

Published by The Royal Society of Chemistry,
Thomas Graham House, Science Park, Milton Road,
Cambridge CB4 0WF, UK

Registered Charity Number 207890

For further information see our web site at www.rsc.org

Preface

Food is a fundamental requirement for all living beings on Earth and it is everyone's right to have their fair share. It is surprising that in the 21st century we still have around 815 million people that are undernourished and 16,000 children die every day due to hunger-related causes. A report by the World Institute states that the richest 1% of adults owned 40% of the world's assets in the year 2000. While such disparities exist, it will be difficult to alleviate food poverty in the world; however, as individuals we can all do our bit.

In Europe alone we waste 30% of food we produce, which is alarming when we know that this material could be utilised elsewhere. Domestic waste could be reduced by composting food waste and there has been a major effort in the UK to instigate such activities. Aside from domestic waste, there is a huge opportunity to utilise industrial food waste and this book focuses on this very matter. Industrial food waste tends to be collected in a controlled way and normally has some low-value use. With growing demand on petrochemical feed stocks it has been evident from recent price fluctuations that this is not a reliable and sustainable carbon source for the future. The spikes in crude oil prices observed in 2008 caused global chaos and mass food shortages of staple materials such as rice and corn in poverty-stricken countries like Bangladesh. It would be wrong to accredit such problems just to the petrochemical economy because it is well known that the extensive biofuels program in the USA also had a direct impact on global food supplies. Biofuels are a complicated subject because the sustainability impact depends on so many things. For example, where there is ample land and water, such as in Brazil, biofuels can be produced sustainably without affecting the local agricultural requirements. However, in other places like India where much attention has been given to plants like Jatropha that can grow on brownfield land without much effort, it is not sustainable to produce biofuels because water is diverted from agricultural crops. To avoid such issues in this book, we decided to focus on side streams and wastes from the food

RSC Green Chemistry No. 24
The Economic Utilisation of Food Co-Products
Edited by Abbas Kazmi and Peter Shuttleworth
© The Royal Society of Chemistry 2013
Published by the Royal Society of Chemistry, www.rsc.org

supply chain. We need to treat such side streams from a biorefinery perspective where the word "waste" does not exist. The biorefinery concept is based on the utilisation of a wide variety of biobased raw materials that are processed using advanced, efficient and clean technologies that generate food, chemicals, materials and energy for society.

The book starts with an introduction to the biorefinery concept and Green Chemistry. As food is so critical to human society, and is mainly controlled by governments, Chapter 2 is dedicated to international food waste policy. In terms of converting food waste into valuable chemicals, we have chosen a state-of-the-art process that we believe is one of the most promising technologies of the future; microwave pyrolysis. Thermochemical conversion of biomass is a simple and efficient method of obtaining high-value chemicals from any type of biomass and Chapter 3 is dedicated to research developments in this field. The book then focuses on end products derived from food waste and their industrial applications. For example, many chemicals derived from food waste can be efficiently used in the environmental and agricultural fields as discussed in Chapter 4. Waste starch from the food manufacturing industries is rich in polymer networks and can be used as a backbone for many adhesives, latexes and coatings and this is discussed in great detail in Chapter 5. Another waste from the food industry is waste cooking oil that can have huge environmental problems as it is not readily biodegradable. Chapter 6 discussed key chemistries that could be employed to convert cooking oil into biofuels but also valorise any side streams such as glycerol. Chapter 7 focuses on the mechanical conversion of agricultural byproducts such as rapeseed meal and sunflower cake into biodegradable materials.

We would like to personally thank all the authors who have contributed to this book and wish them fortune in their personal and work-related goals. It has been a great pleasure to work with some of the worlds leading academics and we hope this book makes a beneficial contribution to science and society.

Dr Abbas Kazmi
University of York

Contents

RSC Green Chemistry No. 24
The Economic Utilisation of Food Co-Products
Edited by Abbas Kazmi and Peter Shuttleworth
© The Royal Society of Chemistry 2013
Published by the Royal Society of Chemistry, www.rsc.org

**Chapter 6   Food Waste and Catering Waste; Focus on Valorisation
of Used Cooking Oil and Recovered Triglycerides   130**
Lorenzo Herrero Dávila

CHAPTER 1

Green Chemistry and the Biorefinery

ABBAS KAZMI

University of York, York, UK
Email: sk_abbaskazmi@yahoo.com

1.1 Introduction to the Biorefinery Concept

The core component of all biorefinery definitions is the conversion of biomass into several products (materials, chemicals, energy, food and feed) and the integration of various technologies and processes in the most sustainable way. The definition developed by the International Energy Agency (IEA) Bioenergy Task 42 Biorefineries has been widely accepted due to its general and broad character:

"Biorefining is the sustainable processing of biomass into a spectrum of marketable products and energy"

This definition includes the following key words:

- biorefinery: concepts, facilities, processes, cluster of industries;
- sustainable: maximising economics, minimising environmental aspects, fossil-fuel replacement, socioeconomic aspects taken into account;
- processing: upstream processing, transformation, fractionation, thermo-chemical and/or biochemical conversion, extraction, separation, downstream processing;

RSC Green Chemistry No. 24
The Economic Utilisation of Food Co-Products
Edited by Abbas Kazmi and Peter Shuttleworth
© The Royal Society of Chemistry 2013
Published by the Royal Society of Chemistry, www.rsc.org

- biomass: crops, organic residues, agroresidues, forest residues, wood, aquatic biomass;
- spectrum: more than one;
- marketable: a market (acceptable volumes and prices) already exists or is expected to be developed in the near future;
- products: both intermediates and final products, *i.e.* food, feed, materials, and chemicals;
- energy: fuels, power, heat.

Using biomass as a sustainable renewable resource is the only way to replace carbon from fossil sources for the production of the carbon-based products such as chemicals, materials and liquid fuels.

In order to be competitive with crude-oil-based products, an integrated biorefinery strategy has been developed to optimise the added value from biomass. This strategy is mainly based on the transfer of petroleum refineries logic to biomass (raw material fractionation, integration of mass and energy fluxes; integration of processes) in order to be able to produce a spectrum of products and therefore maximising the added value. The approach requires the valorisation of the whole biomass. In other words, a biorefinery concept is based on a zero-waste concept.

Moreover, the biorefinery concept goes beyond the petroleum refineries logic, as it includes the management of sustainability based on a cycle concept. This is obvious for renewable carbon at global scale. The cycle also concerns water and mineral nutrients at the local scale, especially nitrogen, phosphorus and potassium (NPK). Contrary to carbon, these elements have to be left on or reincorporated into the soil to avoid depletion, and thus the use of fossil-based fertilisers to compensate for that soil depletion. More generally, the biorefinery concept includes the management of all sustainability issues, including environmental, economic and societal factors.

According to the project *Biorefinery Euroview*,[1] "Biorefineries could be described as integrated biobased industries using a variety of technologies to make products such as chemicals, biofuels, food and feed ingredients, biomaterials, fibres and heat and power, aiming at maximising the added value along the three pillars of sustainability (Environment, Economy and Society)".

All biorefineries are biomass-based industries, whereas not all biomass processing plants are biorefineries. It is important to clarify the respective differences in the next section, in order to understand the focus of this biorefinery vision document.

In conventional biomass processing plants, biomass is directly transformed (1st conversion) into a single main product (usually already marketable). In a biorefinery, however, raw products are firstly converted into intermediate products (1st conversion), which are partly or entirely preproducts. These are further processed (2nd conversion) to several end-products or semifinished goods by additional conversion and conditioning steps, predominantly at the same location.

The additional conversion and conditioning steps are carried out to achieve a better valorisation of biomass by transforming the raw product(s) as completely as possible into various value-added end-products.

For a better differentiation of biorefineries, the following listing provides examples of biomass processing plants that are **not** considered to be biorefineries;[2]

- Plants for biomass conversion that convert the feedstock into one quantitatively dominating, marketable product directly after the primary refining step. Examples are biodiesel plants (main product: biodiesel) or agricultural biogas plants (main product: bioenergy, namely power and heat).
- Plants for biomass conversion that have no combined primary and secondary refining step at the same location. Examples are paper mills without connected pulp mills, separate fermentation plants or starch mills without connected conditioning processes.
- Plants for biomass conversion, where the biomass compounds are not separated, but unmodified or only slightly modified biomass is used or processed. Examples are wood-processing saw mills, or plants producing natural fibre insulation.

A 2030 vision for biorefineries was developed during the FP7 Star-Colibri project that involved European Technology Platforms, Industry Leaders and world-leading academic centres such as the Green Chemistry Centre of Excellence (University of York). Although the work was done in great detail, summaries can be found of this and the 2020 research road map on the project website (www.star-colibri.eu/).

1.1.1 Integration with Existing Industrial Value Chains or Development of New Value Chains

In 2030 many biorefineries will operate at a large-scale commercial level. Most of these biorefineries will be developed based on the integration with existing industrial value chains (top-down approach).

Different biorefineries will be developed based on industrial specificities (sector types) or on geographical specificities (biomass type, quality and availability, infrastructure, presence of a certain industry, *etc.*). The choice of the technological options (processes, feedstocks, location and scale) within the biorefinery will be made by the industrial actors, based on their competitive advantages (available industrial equipment, technological and industrial know-how, access to biomass). Biorefinery development will be driven by the industrial leaders from sectors such as agroindustry, forest-based industry, energy sector (power and heat), (bio)fuels industry and chemicals.

However, another interesting development path for biorefineries is envisaged on the development of new industrial value chains (bottom-up approach). This refers to newly developed, highly integrated, zero waste sites to produce a broad variety of products for different markets from different, pretreated and preseparated (lignocellulosic) biomass fractions. Usually, the whole biomass crop is used (*e.g.* woody lignocellulose, grains and straw from cereals, green grass). In 2011 this approach is often still only operating at pilot or demonstration stage (*e.g.* a lignocellulosic biorefinery in Leuna, Germany). However, a lot of research and development will lead to implementation on a commercial scale in 2030. Preferably these biorefinery plants for new industrial value chains should be integrated in an already existing industrial park to profit from the infrastructure.

In any case, the sustainability and the competitiveness of the different value chains will always rely on a close collaboration within industry sectors and also on a high level of integration between the different production processes.

1.1.2   Biorefinery Scale

The choice of the optimal biorefinery scale has to accommodate the constraints that arise from logistics, production costs and processes. The chosen scale will have a major impact on the emergence of industrial biorefineries and their distribution;

- Large-scale integrated biorefineries, mainly based on thermochemical process, are likely to emerge in Northern Europe and/or in industrial harbours.
- Small/medium-scale integrated biorefineries, mainly based on biotech processes, are likely to emerge in rural areas in "mid" Europe (western, central and eastern Europe).
- Decentralised biorefineries will also emerge in both regions, based on the development of a network of pretreatment units.

As a consequence, the scale has a major impact on the technology choice and on the industrial strategies as it could limit the size of the production facilities (limited biomass quantity per industrial unit). Basically, three possibilities are offered:

- small/medium-sized production facility;
- medium-sized/large production facility linked to a network of decentralised biorefineries (biomass fractioning and/or concentrating units);
- very large production facility, located on industrial harbours with importation of biomass.

1.1.3   Biomass Supply: Harbour (Import of Biomass or Intermediates) or Rural (Locally Produced Biomass)

As a consequence of the biomass supply form, there will be not one but several biorefinery types in Europe, with a predominance of certain types according to the geographical biomass location.

- The biorefineries based on wood (locally produced biomass) are likely to be developed in Northern Europe or in dense forested area in "mid-Europe".

- The biorefineries based on classical agricultural crops (cereal, sugarbeets, oilseed crops) are likely to be developed in "mid-Europe".
- Biorefineries based on imported biomass will be established mainly in or very near to large harbours (like Rotterdam).
- The development of biorefineries in South Europe is more difficult to predict. It could be either connected to the area of industrial harbours or to (new) regional crops.

The BIOPOL (2009) project gave some predictions about the most likely regions for biorefinery development in connection to the biomass availability. The main conclusions were: "Western Europe has the best prospects for biorefinery development. It has: high agricultural yields, vast amounts of lignocellulosic agricultural side streams, considerable forestry and good possibilities to sell biorefinery side products. The countries in the East of Europe have good opportunities to improve agricultural yields. Thus, they could become interesting countries for biorefinery establishment. Northern Europe is currently a natural market leader of lignocellulosic biorefinery due to the presence of large forests."

1.1.4 Biorefinery Concepts in 2030

Some more traditional biorefinery concepts were already established on an industrial scale in 2011. They are based on an extension and/or on upgrading processes of existing industrial plants in the respective sectors. However, there will emerge other, newly developed biorefinery concepts that will be well established in 2030. In 2011, these biorefinery concepts were still only in the research, development or demonstration stage. For some of the following future biorefinery concepts the first pilot plants are being built in Europe in 2011.

1.1.4.1 Starch and Sugar Biorefineries

Starch and sugar agroindustries have a long experience in starch fractioning and/or fermentation and distillation. They are therefore a perfect candidate to integrate biotech processes for first- and second-generation bioethanol and, in a second step, other fermentation products. Starting from production based on starch and sugar crops, the industrial units will progressively use lignocellulosic feedstocks and integrate fractionation processes. The first steps will be the integration into the supply chain of cereal straw and, in a second step, of dedicated lignocellulosic (mainly herbaceous) crops.

The integration scenario does not concern only biomass diversification but also the valorisation of side products of the lignocellulose deconstruction: lignin, C5-sugars from hemicellulose and C6-sugars mainly from cellulose. The utilisation of lignin as an energy source by cogeneration will be progressively replaced by the development of new chemistry based on lignin. The ethanol production from C5-sugars, on account of the poor conversion yield, will be replaced by the development of a new C5-chemistry (by biotech and/or chemical processes) to produce higher-value chemicals (2020–2025 horizon). The ethanol production from C6-sugars will be progressively replaced by new fermentation processes and the production of higher-value chemicals. This development is likely to occur several years after the C5 switch (2025–2030) as the bioethanol European market will still be growing until this period.

The required biomass quantities per biorefinery are in the range of 200 to 400 kt/year of dry biomass, which enables reliance on a local production area, and the integration of the management of sustainability parameters in the production chain (carbon sequestration, nitrogen and other mineral nutrients cycles).

This will lead to the development of small/medium-scale rural biorefineries close to the agricultural production areas producing the required biomass. These rural 'starch and sugar' biorefineries will be implanted in the most efficient production and supply areas. Ideal localisation will be "mid-Europe" (from West to East Europe).

1.1.4.2 Oilseed Biorefineries

The oilseed agroindustry will develop different strategies. Today, it focuses on first-generation biodiesel and the development of the oilseed-based biorefineries could involve the integration of the glycerol valorisation and the development of a glycerol-based chemistry. However, the main evolution of these biorefineries will be based on the development of a new oleochemistry, based on long-chain fatty acids from European oilseeds (mainly rapeseed and sunflower) and the progressive integration of oleochemical processes into the biodiesel production chain. Moreover, this shift will be supported by the evolution of biofuels production in Europe and the relative decrease of first-generation biodiesel on account of its low energy efficiency per surface unit.

1.1.4.3 Forest-Based (Pulp and Paper) Biorefineries

The forest-based (pulp and paper) industry is located close to the main forest areas in Europe (mainly in Northern Europe). The industry has a long experience of woody and lignocellulosic biomass logistic. Wood and pulp byproducts (such as bark) are relatively dry biomass and are therefore well suited for new thermochemical conversion processes such as gasification. The industry is therefore a good candidate to integrate advanced second generation biofuels production and/or chemicals production from syngas (*e.g.* DME). The industry has also access to a huge amount of lignin, which is currently mainly used for combustion to produce bioenergy. Higher added-value chemicals will be obtained by integrating a chemical valorisation within forest-based biorefineries, particularly focusing on black liquor.

1.1.4.4 Biofuel-Driven Biorefineries

In 2011, there is no industry in Europe yet that has biorefineries using gasification in combination with the Fischer–Tropsch process to produce liquid second-generation biofuels. Since huge investments are needed to set up a large-scale industrial unit, the best candidates will be either energy companies or traditional oil companies, because of the infrastructure availability and the economy of scale. In 2030 oil companies will have installed large-scale biorefineries based on thermochemical process into their existing oil refineries located in the main European harbours. The required biomass (wood, forest residues, dedicated lignocellulosic crops or urban wastes) will be imported and also collected locally. Another interest of the traditional oil companies will be the utilisation of hydrogen from syngas as a source for hydrogenation of heavy crude oil. The gasification units will also be used to produce higher-value chemicals by catalysis processes.

1.1.4.5 Green Biorefinery

A Green Biorefinery processes wet biomass, such as grass, clover, lucerne and alfalfa (BIOPOL, 2009). The wet biomass is pressed to obtain two separate products: fibre-rich press juice and nutrient-rich press cake. The press cake fibres can be utilised as green feed pellets or as a raw material for chemicals. The press juice contains valuable compounds, such as proteins, free amino acids, organic acids, minerals, hormones and enzymes. Lactic acid and its derivatives as well as ethanol, proteins and amino acids are the most favourable end-products from press juice.[3] The bio-organic residues in press juice are mainly used to produce biogas with subsequent generation of heat and electricity.

An example in 2011 of a pilot of this future biorefinery concept is the production and demonstration plant of the Biowert GmbH in Brensbach, Germany, where insulating material, reinforced composites for production of plastics and biogas for heat and power are generated from grass in an inte-grative process. In 2030 many of these smaller-scale Green Biorefineries will be established in regions that traditionally produce high quantities of wet biomass (like grassland areas).

1.1.4.6 Future Lignocellulosic Biorefineries

The lignocellulosic biorefinery concept based on dry biomass is not only applicable for previously described pulping process. Two different approaches can be distinguished in 2030 for the lignocellulosic biorefinery: thermochemical and biochemical.

The **thermochemical** approach is based on gasification of lignocellulosic feedstocks, and further processing the syngas to transportation fuels and chemicals. Many different biomass types are taken into consideration as raw material for this type of lignocellulosic biorefinery concept: dry agricultural residues (*e.g.* straw, peelings, husks), wood, woody biomass, and biogenic residues (*e.g.* waste paper, lignin).

Another option is the **biochemical** approach, that is based on the biochemical fractionation of the lignocellulosic raw material (cellulosic agricultural residues, forest residues or dry biogenic waste materials) into three separate product precursors: cellulose, hemicellulose and lignin. This is done during the primary refining step. These chemical fractions are then treated separately, and converted into value-added products during the secondary refining step. Cellulose can be hydrolysed into sugars and then used as fermentation substrate to produce alcohols (*e.g.* ethanol), organic acids and solvents. The second-fraction hemicellulose can be converted to xylose, gelling agents, barriers, furfural and further to nylon. Finally, lignin can be applied as a binder and adhesive or can be used for the production of fuels, carbon fibres for materials or syngas, which can be used for energetic or industrial purposes.

The thermochemical (gasification) approach will be mainly based on woody biomass (forest biomass and residues or dry biogenic waste materials), while agricultural residues (*e.g.* cereal straw) and dedicated lignocellulosic herbaceous crops will be more likely used in the biochemical fractionation and hydrolysis processes. We assume that the technical problems in the biochemical approach to overcome the "biomass recalcitrance" will be solved in the period 2011–2020, and that both approaches will lead to commercially viable lignocellulosic biorefineries in 2030.

1.1.4.7 Aquatic (Marine) Biorefinery

Aquatic (marine) biomass (microalgae and seaweeds) is an interesting new feedstock for aquatic biorefinery, characterised by high productivity per area unit, and a high content of valuable components for the biobased economy, including: oils, proteins, polysaccharides and specific biomolecules. Aquatic biomass cultivation and processing are regionally or locally integrated. Due to the varied composition, microalgae and seaweed biomass are highly suited for biorefinery with end-products ranging from fuels and bulk chemicals to specialty chemicals and food and feed ingredients.

Microalgae, like some other micro-organisms and plants, produce storage lipids in the form of triacyglycerols (TAGs) that can be used to synthesise fatty acid methyl esters (a substitute for fossil-derived diesel fuel). Microalgae represent a very attractive alternative compared to terrestrial oleaginous species, because their productivity is much higher, and they do not compete for land suitable for agricultural purposes, providing therefore food security. Attractive features of microalgae include: high areal productivity as compared to terrestrial crops, location of cultivation systems on marginal lands or other low-grade surfaces, the unique biomass composition including the ability to accumulate large amounts of oils, and the great variety in species and products. Furthermore, algal cultures lend themselves for combination with wastewater treatment, CO_2 removal from flue gas, and useful applications for low-temperature waste heat.

Typical products from microalgae include:

- oils for food applications and nutraceuticals such as omega fatty acids;
- oils and derived fractions for nonfood applications, *i.e.* transport fuels, chemicals;

- proteins and derived products (amino acids, N-chemicals);
- other bulk chemicals (*e.g.* ingredients for coatings);
- a range of high-value specialties.

In 2030 several commercial biorefinery concepts for integral microalgae-based production chains will exist for:

- biofuels and coproducts;
- feed for aquaculture, including recycling of nutrients;
- production of food or food ingredients and coproducts.

The cultivation of seaweed can be an enormous source of biomass in the future. Use of land for cultivation of biomass for the production of fuels, products and fibres is subject to debate due to the competition with food production. Expansion of biomass cultivation to the sea would increase the potential amount of biomass being available for nonfood and nonfeed purposes, and thereby increasing the potential share of biomass in renewable energy supply in the future.

For aquatic biomass several biorefinery processes seem to fit. Depending on the type of feedstock (the seaweed species), and the possibilities of process integration, a processing route will be selected. Hydrothermal upgrading (HTU), anaerobic digestion to methane, and ethanol fermentation and distillation, are possible processes.

1.2 Extractable High-Value Chemicals in Food and Byproducts

Secondary metabolites are extracted from numerous plants and flowers and are used as pigments, health products, food ingredients and cosmetic applications. This section focuses on mainly European crops. The biomass has been divided into seven distinct groups that are cereal, oilseed, root, wood, herbaceous and marine biomass.

The bulk of secondary metabolites are used in the food, cosmetics and health industries and good sources of information are available on the internet. For example, hundreds of secondary metabolites are used in the cosmetics industry and in the food and health industry.

1.2.1 Cereal Crops

Human society has depended on cereal crops for millennia and today there are thousands of products based on wheat, barley, oats and rice. Although much of the focus is on the grain for food purposes, there are very important metabolites that are not used in the growth of the plant, however when extracted, they have many uses in various applications. Applications exist in the cosmetics industry where extracts are used in products such as hair conditioner, shampoo, bleacher, styling gel/lotion, antiageing cream, facial moisturiser, hair spray, body wash, foundation and mascara.

Secondary metabolites mainly exist in the straws of the crops in the form of waxes. However these are only 1–3% by weight of the straw, therefore any components that are to be extracted must be of very high value for the process to be economical. For example, the composition of wheat straw wax contains several commodity chemicals, however, the likelihood of using this as a feedstock are low. A list of wheat straw extracts is shown in Table 1.1.

When a low-cost method of extracting such waxes is developed then the components could have significant commercial value. For example, the waxes can be used in the lucrative cosmetics industry for applications such as lipstick.[4] The waxes also contain a series of chemicals that act as insect repellents and therefore could have commercial value in this field. However, the most valuable components of the waxes are likely to be the sterols and polycosanols that have been shown to act as cholesterol-reducing agents. Polycosanols are widely used in the world in the form of pills and are commercially extracted from

Table 1.1 Extracts from wheat straw and their markets.

Product	Market	Volume	Current Feedstock
Cetearyl Alcohol	Personal Care	Large	Crude Oil and coconut oil
Benzoic acid	Food preservatives, Chemical precursor	139,000 tonnes +	Crude Oil
Succinic acid	Chemical intermediate, Numerous applications	Large	Biomass *via* fermentation
Fumaric acid	Medicine, Food, Chemicals	Large	Nonbiomass
rn-Toluic acid	Insect repellent, PVC Stabiliser	Low	Nonbiomass
Salicylic acid	Medicine, Cosmetics	High	Biomass
Maleic acid	Cosmetics	Low	Nonbiomass
p-Hydroxybenzoic acid	Cosmetics, Chemical intermediate	–	Nonbiomass
Gentisic acid	Mass Spectrometry, Medicine	Low	Nonbiomass
Vanillic acid	Flavouring agent, Chemical intermediate	–	Biomass
p-Resorcylic acid	Cosmetics and fine chemicals	–	–
Protocatechuic acid	Antioxidant	–	–
Azelaic acid	Medicine	–	Biomass
1,2,3,5-Tetrabromobenzene	–	–	–
Dihydroferulic acid	–	–	–
trans-p-Coumaric acid	Antioxidant	–	–
Syringic acid	–	–	–
cis-Ferulic acid	–	–	–
trans-Ferulic acid	–	–	–
1-Naphthoic acid	–	–	–

sugar-cane waxes. Furthermore the wheat straw contains a relatively high amount polycosanols[5] (164 mg/kg) which makes it an attractive component to extract. Sterols also exhibit similar health benefits and are currently extracted from vegetable oil and pinewood, which do not meet the ever increasing demand.[6] Although polycosanols and sterols employ different mechanisms to reduce cholesterol levels, mixtures of both extracts could offer enhanced effectiveness and this would mean more of the wheat straw wax is used.

The use of sterols in food products has also been approved by the EU Novel Food program in 2004, where cholesterol reducing ingredients can be used in a wide variety of products such as margarine, milk, yoghurt, cheese, soymilk, dressings and rye bread.[7] The sterols are now being used in one-shot drinks and special milks with the market still growing. Large companies such as Unilever have taken the lead, however, smaller companies are also entering this lucrative market. Although in Europe there are some restrictions for use in some foods, in the US there are no restrictions and sterols are generally regarded as safe. A report from Frost & Sullivan valued the European plant sterols market at $185 million and this is estimated to rise to $395 million in 2012, see Figure 1.1.

1.2.2 Oat Extracts

Sterols also exist in oats, however, oats also have a considerable amount of antioxidants such as tocols (vitamin E), phytic acid, phenolics and avenanthramides.[8] These important components are concentrated on the outer layers of the kernel and although they are consumed as part of the grain, they can be extracted using solvents such as methanol or choloroform/propanol mixtures to be used in food supplements or blended with other food products.

Figure 1.1 Sterols from wheat straw wax could also be used in consumer health products.

Table 1.2 The concentration of tocols found in oats from different parts of the world.

Region	Concentration of tocol (mg/kg)
USA	19–30
Hungary	15–48
UK	19

The tocols found in oats are mainly in the form of α-tocotrienol and concentrations vary depending on genotype and location as shown in Table 1.2.

On average, barley contains twice as much tocols as oats therefore it is also a good source for such molecules. Phenolic acids are found in oats in the form of ferulic, vanillic, sinapic, ρ-coumaric and ρ-hydroxybenzoic acid. Clearly the concentrations of these acids vary depending on the species and the section of the plant that they are extracted from. Table 1.3 shows that high concentrations of ferulic acid that can be obtained from the groats and hulls, whilst high concentrations of vanillin, hydroxybenzoic and ρ-coumaric acid can also be found in hulls.

Cholesterol-reducing medication is also available such as LIPITOR, which had sales of $12 billion in 2008. Synthetic antioxidants such as butylated hydroxytoluene (BHT) and butylated hydroxyanisole (BHA) are dominant in this market, however, these synthetic molecules have some key disadvantages as they are suspected to have toxic properties, are highly volatile and restricted by certain countries.

1.2.3 Oilseed Extracts

Oilseeds residues contain significant amounts of phenolics, falvanoids and sinapine that can be used as natural antioxidants. Furthermore, they can add value to residues that normally would be used as animal feed. Using conventional solvents such as methanol, acetone, water, and ethyl acetate/water these components can be extracted with yields of 3 to 19% of all extracted material.[9] As shown in Table 1.4, significant amounts of these valuable antioxidants are found in various oil seed crops.

The oilseed cake is a material that is left after the oil has been extracted from the seeds and contains significant levels of protein and chemicals that have numerous applications. Currently, there exists a stable, but low-value animal feed market for oilseed cake. However after extraction of nutrients and proteins, the cake will no longer be viable for animal feed and the only application would be as fuel for energy production, also a low-value market. The Sustoil project, funded by the EU under the 7th framework programme, has identified a number of valuable extracts that can be obtained from oil seed rape cake.[10]

Table 1.3 Phenolic acids, alcohols, and aldehydes concentrations in oat groats and hulls (mg/kg).[8]

Compound	Groats, free acid fraction				Groats, soluble esters	Groats, insoluble bound	Hulls, free acid	
	Sosulski et al.[19]	Xing & White[20]	Dimberg et al.[21]	Emmons & Peterson[22]	Sosulski et al.[19]	Sosulski et al.[19]	Xing & White[20]	Emmons & Peterson[22]
Caffeic acid	1.0	16.8	2.2	2.4		1.6	0.1	0.9
Catechol		tr						
Coniferyl alcohol			0.8					
o-Coumaric acid							6.9	
p-Coumaric acid	0.7	44.9	1.6	0.9	0.5	0.8	59.7	9.7
Gallic acid				1.3				0.6
Ferulic acid	2.4	147.2	2.3	1.2	8.6	55.3	142.3	1.7
p-Hydroxybenzoic acid	0.7	3.5			0.7	tr	50.0	
p-Hydroxybenzaldehyde			0.9	0.3	tr			
p-Hydroxyphenylacetic acid	0.4	0.6			tr		4.6	7.7
Protocatechuic acid	0.5			0.7		tr		2.1
Salicylic acid	tr					tr	3.1	
Sinapic acid	2.3			0.5	4.3	tr	5.6	0.6
Syringic acid					3.0	tr		
Vanillic acid	0.7	16.1	1.2	1.6	3.5		24.3	4.0
Vanillin		3.4	2.3	1.0			54.2	6.3

Table 1.4 Total extractable compounds (EC), total phenolic compounds
(PC), flavanoids, and sinapine of different extracts obtained from
fat-free residues of oilseeds (results are given in mg/g of extract).[9]

	EC (mg/g)	*PC (mg/g)*	*PC/EC (%)*	*flavanoids (mg/g)*	*sinapine (mg/g)*
		70% methanol extract			
B. carinata	223.0	13.6	6.1	7.49	99.9
rapeseed	187.3	11.8	6.3	7.96	94.6
L. campestre	311.0	35.6	11.5	128.80	0.5
B. Vema	159.0	13.1	8.2	11.47	100
crambe	214.0	8.1	3.8	9.36	9.6
sunflower	102.0	16.1	15.8	12.03	0.0
C. sativa	166.7	11.1	6.6	142.79	56.5
mustard	245.0	17.6	7.2	0.97	122.1
		70% acetone extract			
B. carinata	192.7	9.4	4.9	9.32	129.7
rapeseed	143.3	6.6	4.6	5.20	85.6
L. campestre	150.7	7.4	4.9	72.49	1.8
B. Vema	156.7	8.5	5.4	18.61	100
crambe	224.0	9.6	4.3	5.69	11.5
sunflower	110.0	4.7	4.3	11.23	0.0
C. sativa	309.3	11.1	3.6	36.15	47.5
mustard	228.0	17.6	7.7	1.21	85.5
		water extract			
B. carinata	466.7	36.0	7.7	2.04	17.5
rapeseed	326.0	21.3	6.5	0.74	11.7
L. campestre	229.3	44.7	19.5	3.54	0.7
B. Vema	188.7	25.2	13.4	5.26	57.0
crambe	332.7	16.6	5.0	0.64	2.6
sunflower	267.3	38.8	14.5	1.45	0.0
C. sativa	275.3	21.8	7.9	11.77	9.6
mustard	247.3	36.5	14.8	0.07	32.5
		ethyl acetate extract			
B. carinata	47.3	3.6	7.7	8.25	37.9
rapeseed	72.0	4.0	5.5	4.19	39.4
L. campestre	59.3	8.3	13.9	47.92	0.0
B. Vema	60.0	5.4	9.0	19.83	0.0
crambe	74.0	2.6	3.6	9.30	13.7
sunflower	80.7	2.7	3.3	4.49	0.0
C. sativa	58.0	3.2	5.5	59.75	0.0
mustard	60.7	9.2	15.2	2.18	60.9

For example, glucosinolates are an important group of chemicals found in oil
seed rape, turnip, broccoli and mustard that can be broken down by enzymes to
produce isothiocyanates that have pesticidal properties. There is great potential
to use such pesticides in organic farming. An even greater market exists in the
food sector as it has been shown that glucosinolates offer anti-inflammatory,
antimicrobial and chemopreventive effects. Oil seed rape also contains
significant amounts of phenols in the form of sinapine, which is present at
around 1% by weight in the cake.

1.2.4 Root Crops

Sugarbeet pulp contains a variety of valuable components, most notably pectin, a complex polysaccharide consisting of D-galacturonic acid and a series of neutral sugars such as L-rhamnose, L-arabinose, and D-galactose.[11] Pectins are used widely in the food industry as gelling agents in jams and are mainly sourced from apple pomace and citrus peels. Sugarbeet is also a potential source as it contains a very high amount of pectin, reported to be 15–30% of the dry weight, and its pectin also offers superior emulsifying properties, however, poor gelling is observed. Pectin can also be extracted from sunflower residues at 15–25% dry weight.

The global pectin market stands at 30,000 tonnes and the price ranges from $11–$13/kg, see Table 1.5.[12] The San-Ei Gen F.F.I. Inc company has recently developed a process that modifies sugarbeet pectin to offer enhanced emulsifying properties and is close to commercialisation.[13]

1.2.5 Wood and Bark

Tannins, or wood polyphenols, are important extracts from wood and bark and have been used in various products such as leather. Other applications for tannins include use in particleboard adhesives and anticorrosive primer. A lucrative market for tannins is developing in the field of medicine where they have shown to have anti-inflammatory, antiviral, antimicrobial and anti-parasitic effects.

Another important component called oleoresin is found in pine and other softwood. The oleoresin, isolated by tapping of living trees, is fractionated into gum rosin and gum turpentine. Approximately 75% of the global oleoresin production (over 1,000,000 tons) and processing takes place in China. Rosin is used in numerous applications ranging from industrial inks to soaps. Rosin can be used as a glazing agent and is used in medicine and chewing gum. A small percentage of rosin is also in flux, required for soldering. Other valuable components of wood that have great potential include phytosterols, flovonoids

Table 1.5 The growth in the use of pectin in various food products.

Application	Growth Rate
Fruit preparations	15–20%
Fruit spreads (low-sugar jams)	10% +
Yoghurt and dairy	15–20%
Acidified milk drinks, high-sugar jams	2–5%
Fruit juices and high-calcium drinks	10–15%
Baked goods, confectionary	5%
Fat replacers	20% +

(*e.g.* Pycnogenol), lignans and other bioactive substances, and other specialty chemicals.

In addition, various hemicelluloses can be isolated from wood for the applications in their polymeric or monomweric forms, after hydrolysis. For example, some arabinogalactan isolated from larch wood by hot-water extraction has been on the markets since the 1970s.

1.2.6   Herbaceous

Grasses are an abundant source of biomass and can be grown in many areas throughout the world. When grasses are pressed a green juice is produced that contains a cocktail of valuable substances. The juice can contain hundreds of individual components that fit into the following groups; proteins, lipids, glycoproteins, lectins, sugars, free amino acids, dyes, hormones, enzymes, minerals and others.[14]

Sugars such as glucose, fructose, fructan, erythrose, rhamnose, xylose, galactose, mannose, mannitol and maltose are contained within the juice and have considerable value. Sugars can be converted to high-value chemicals and can therefore penetrate several markets from fuels to cosmetics. Beta-carotene and xanthophyll are possible anticarcinogens and used for applications in cosmetics, food, textiles and toys. The juice also contains other vitamins such as B1, B2 and E that all have considerable market value. The juice also contains high-value fatty acids such as palmitic acid, linoleic acid and linolenic acid that not only provide health benefits but can also be used in the cosmetics industry. The proteins within the juice are of high value as they can be used in medical diets to promote recovery from brain damage and can be consumed by people with kidney problems.[15]

1.2.7   Summary

Secondary metabolites exist in almost all types of biomass, as shown above, and are mainly used in speciality markets such as food ingredients, cosmetics and health products. There are certain biomass types that have not been covered in this study but are promising alternatives. In order for a secondary metabolite to be used in any product its production cost must be low. In exceptional cases where the extract is of very high value then costly processing may be utilised but in most cases a low-cost production method is required. For example, wheat straw is already used as a low-value product in animal feed and for energy production, however, extracting valuable components should not increase the cost to such an extent that the straw is overpriced. Alternatively, there are many biomass sources such as certain fruits and vegetables that are already being used for consumption purposes and also contain high-value secondary metabolites. In Table 1.6 a list of important metabolites found in various types of fruit and vegetables that have potential health benefits are shown.

Table 1.6 A selected list of phytochemicals found in various biomass.[16]

Name	Source	Uses
Terpenoids /Isoprenoids		
Carotenoid Terpenoids		
Lycopene	Tomatoes	Cholesterol reducing, suppresses tumour growth, antioxidant
Beta-Carotene	Carrots, Pumpkin	Cornea protection
Alpha-Carotene	Carrots, Pumpkin	Anticarcinogenic
Lutein	Kale, spinach, watercress and parsley	Eye protection
Zeaxanthin	Kale, spinach, watercress and parsley	Eye protection
Astaxanthin	Fish	Powerful antioxidant
Noncarotenoid Terpenoids		
Perillyl Alcohol	Cherries, Mint	Anticarcinogenic
Saponins	Chickpeas, Soybeans	Removes cholesterol, anticolon cancer
Terpeneol	Carrots	Anticarcinogenic
Terpene Limonoids	Orange peel	Anticarcinogenic
Polyphenolics		
Flavonoid Polyphenolics		
Anthocyanins	Blueberries, blackberries, black raspberries	Endothelial cell protection
Catechins	Tea leaf, Chocolate	Anticarcinogenic
Isoflavones	Soy beans	Anticarcinogenic, Cholesterol reducing
Hesperetin	Citrus fruits	Antioxidant, anticarcinogenic
Naringin	Grapefruit	Cholesterol reducing
Rutin	Asparagus, buckwheat, citrus fruits	Anti-inflammatory, Strengthens capillaries
Quercetin	Apple skins, red onion	Antihistaminic, antioxidant, blood thinner
Silymarin	Artichokes, milk thistle	Antiatherosclerotic, antioxidant, anticarcinogenic
Tangeretin	Tangerines	Strong anticarcinogenic
Punicalagin	Pomegranate	Antioxidant, anti-inflammatory
Phenolic Acids		
Ellagic Acid	Raspberries, strawberries	Reduces esophagal and colon cancer
Chlorogenic Acid	Blueberries, tomatoes and bell peppers	Antioxidant
P-Coumaric Acid	Red/Green bell peppers	Antioxidant
Phytic Acid	Legumes, whole grain	Reduces cancer growth, reduces blood-sugar levels
Ferulic Acid	Brown rice, whole wheat, oats	Antioxidant, anticarcinogenic
Vanillin	Vanilla bean	Antioxidant, anti-inflammatory

Table 1.6   (*Continued*)

Name	Source	Uses
Cinnamic Acid	Balsam tree resin, wood, inner bark	Antibacterial, pigment
Other Nonflavonoid Polyphenolics		
Curcumin	Tumeric	Anti-inflammatory, antioxidant, asprin alternative
Resveratrol	Grape skin	Anti-inflammatory
Lignans	Flaxseed, wood	Cytotoxic agent
Glucosinolates		
Isothiocyanates		
Phenethyl Isothiocyanate	Watercress	Anticarcinogenic
Benzyl Isothiocyanate	Cruciferous plants	Anticarcinogenic
Sulforaphane	Broccoli	Anticarcinogenic
Indoles		
Indole-3-Carbinol	Broccoli	Anticarcinogenic
Thiosulfonates	Garlic, onions	Reduces blood pressure
Phytosterols		
Beta-Sitosterol	Black cumin seed, cashew fruit, rice bran	Reduces cholesterol, anticarcinogenic
Anthraquinones		
Senna	Leaves of luguminous herbs	Laxative
Barbaloin	Aloe vera	Laxative

1.3   Transformation of Glycerol into High-Quality Products through Green Chemistry and Biotechnology

Glycerol, or propan-1, 2, 3-triol, is an important byproduct of biodiesel production generated from the transesterification reaction of triglycerides from virgin vegetable oils or fats as well as waste oils, with alcohols including methanol and ethanol, in the presence of a homogeneous base catalyst such as NaOH or KOH, and acid catalyst. In general, the production of 10 kg of biodiesel yields approximately 1 kg of crude glycerol (10% (w/w)), and currently the world's capacity for biodiesel production is dramatically increasing. Further increases in biodiesel production rates will significantly raise the quantity and surplus of crude glycerol and partially purified glycerol in the environment. In contrast to the surplus of impure glycerol, high-purity glycerol is an important industrial feedstock that finds applications in the food, cosmetic and pharmaceutical industries, as well as other more minor uses. However, its refining is generally costly, especially for medium- and small-sized plants.

Glycerol can be used as a building block for many chemicals such as 1, 3-propanediol, lactate and succinate. In fact many companies have initiated commercial plans to manufacture high-value chemicals such as epichlorohydrin

(Solvay SA) and proplylene diol (Ashland/Cargill) from glycerol feedstocks. The market volatility in the price of glycerol has caused concern for these projects, however, the long-term fundamentals remain strong.

Glycerol has been known since 2800 BC mainly as a byproduct of soap production.[17] Currently, glycerol has numerous applications in personal care, food, tobacco, detergents, cellophane, explosives and pharmaceuticals.[18] Leffingwell and Lesser identified 1582 applications for glycerol in 1945.[19] However, in recent times, many glycerol production plants are closing and new plants utilising glycerol as a raw material are starting.[20] Global glycerol production has increased from 60,000 tons in 2001 to 800,000 tons in 2005, partly due to biodiesel production. The amount of glycerol being used in technical applications is around 160,000 tons, and this is expected to grow at a rate of 2.8% per year.[21]

Glycerol is a raw material for the production of flexible foams and rigid polyurethane foams. It is known to provide properties including flexibility, pliability and toughness in surface coatings and paint regenerated cellulose films, meat casings and special quality papers.[22] Glycerol has the ability to absorb moisture from the atmosphere and is therefore used in many adhesives and glues to prevent early drying. In food applications, nontoxic glycerol is used as solvent, sweetener and preservative. Many polyols such as sorbitol, manitol and maltitol are used as sugar-free sweeteners; however, they are facing fierce competition from glycerol. Glycerol has similar sweetness to sucrose and has the same energy as sugar. Furthermore, it does not raise blood sugar levels and does not feed plaque bacteria. Glycerol is also employed as an emollient, humectant and lubricant in many products in the personal care industry including toothpaste, mouthwashes, shaving cream and soaps.[23]

A detailed revision on glycerol transforming processes and applications can be found in 'The Future of Glycerol- New usages for a raw material' authored by Mario Pagliaro and Michele Rossi.[23] The book was published in 2008 and focuses on key chemical and biochemical transformations with detailed processing conditions. In this book, relevant information on the sustainability and economics of glycerol and biofuels production is discussed. The detailed synthetic chemistry involved in the transforming processes has also been reviewed by Behr *et al.* in his paper entitled "Improved utilisation of renewable resources: New important derivatives of glycerol".[24]

A more detailed revision on chemicals that can be derived from glycerol was conducted in 2008 by Zheng, Chen, and Shen in "Commodity Chemicals Derived from Glycerol, an Important Biorefinery Feedstock".[25] Many important chemicals have been identified that can be produced from glycerol-derived platform chemicals and their respective industrial applications are discussed. Furthermore, this review maps the reaction pathways of a glycerol-derived platform chemical that can form many other commodity chemicals that are not easily identifiable. Some of the important commodity chemicals identified include acrolein, dichloropropanol, epichlorohydrin, dihydroxyacetone, 1,3-propanediol, 1,2-propanediol, glycerol carbonate, diacylglycerol (DAG), monoglyceride (MG), oxygenate fuels, glyceric acid, tartronic acid, and mesoxalic acid.

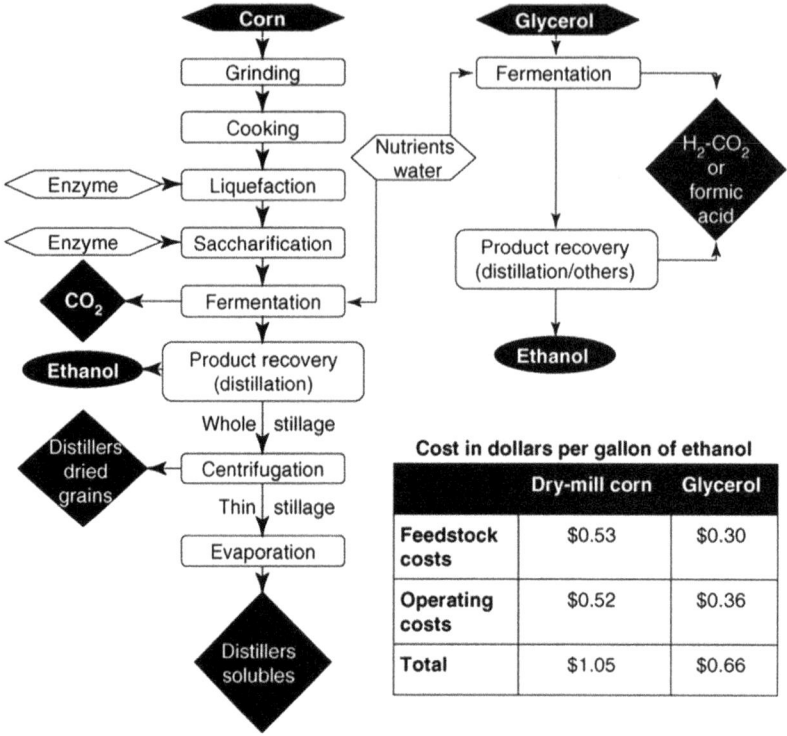

Figure 1.2   Transforming glycerol into commodity chemicals.

Biochemical methods can likewise be employed to transform glycerol into commodity chemicals as highlighted by Yazdani et al.[26] For some transformations a detailed description of the processes involved are shown including the overall production costs as shown in Figure 1.2.

Some of the important commodity chemicals produced using anaerobic fermentation include succinic acid, 1,3 propanediol, propionic acid, formic acid, butanol and ethanol. A recent paper by Silva et al. reviews glycerol as a source for industrial microbiology.[27] This report identifies various microbial reaction pathways to produce many chemicals from glycerol-derived platform chemicals.

1.3.1   Aqueous Phase Reforming

One of the major achievements in glycerol chemistry is the development of aqueous phase reforming processes (APR) that involve the conversion of glycerol to hydrogen and carbon monoxide (synthesis gas). The reported process conditions require 250 °C and the utilisation of a Pt–Re catalyst in a single reactor.[28] This process can also theoretically produce high yields of hydrogen from glycerol at low CO concentrations due to favourable water-gas shift (WGS) thermodynamics. This requires significantly lower energy consumption than traditional methane reforming.

The synthesis gas can be used as a building block for chemicals and fuels *via* Fischer–Tropsch synthesis (FTS). Through FTS, syngas can be converted to a range of useful liquid hydrocarbons (mainly linear alkanes, although alkenes and alcohols can also be produced under certain conditions) using iron and cobalt catalysts. The temperatures used in the process typically range from 150 to 300 °C and pressures of one to a few atmospheres are common. High temperatures lead to gasolines and linear low molecular mass hydrocarbons, whereas lower temperatures and high pressures favour the formation of longer-chain hydrocarbons (*e.g.* waxes).

1.3.2 Selective Reductions

The main processes utilised to reduce glycerol to glycols are hydrogenolysis, dehydroxylation and biotechnology *via* bacteria.

Propylene glycol is commercially produced *via* hydrogenolysis using a copper chromite catalyst at 200 °C at pressures below 10 bar. Wang and coworkers showed that it was possible to produce 1,3 propanediol *via* selective dehydroxylation.[29] The central hydroxyl group of glycerol is selectively converted to a tosyloxy-group that is removed using hydrogenolysis. The biological reduction to 1,3 propanediol involves the use of bacterial strains from groups such as *Citrobacter, Enterobacter, Ilyobacter, Klebsiella, Lactobacillus, Pelobacter* and *Clostridium*. Freund showed in 1881 that PDO could be produced using *Clostridium*, a widely available micro-organism found in nature.[30] The process involves a two-step enzyme-catalysed reaction sequence in which a dehydratase catalyses the conversion of glycerol to 3-hydroxypropionaldehyde, which is subsequently reduced to PDO by a NAD^+-linked oxidoreductase.

1.3.3 Halogenations

The chlorination of glycerol *via* a 1,3-dichloro-2-propanol intermediate yields epichlorohydrin, an important and valuable chemical. 1,3-dichloro-2-propanol can be produced directly from glycerol using HCl as catalyst. Its subsequent dehydrochlorination using NaOH generates epichlorohydrin and NaCl.

1.3.4 Dehydrations

Glycerol dehydration can also produce relevant chemicals including acrolein, 3-hydroxypropionaldehyde and acrylic acid. Protonated glycerol is more susceptible to dehydration due to reduction in the energy barrier of the inter-mediate state. Therefore, acrolein can only be produced in acidic conditions. The reaction can be conducted in both liquid or gas phase at high temperatures and/or vacuum that are normally used to drive the dehydration. In the presence of molecular oxygen, acrylic acid can be produced *via* a one-step oxy-dehydration step.

1.3.5 Etherifications

Glycerol alkyl ethers can be synthesised by etherification of alkenes including isobutylene in the presence of an acid catalyst at temperatures from 50–150 °C. The typical molar ratios used in the reaction are 1 : 2 (glycerol : isobutylene) and the yield can be improved by optimising the reaction conditions. Glycerol can be etherified to form polyglycerol *via* anionic polymerisation of glycidol through a cation exchange equilibrium initiated by partially deprotonated 1,1,1-tris(hydroxymethyl) propane. The resulting polymer usually has a polydispersity of below 1.5 and a molecular weight ranging from 1000 to 3000 gmol^{-1}.

1.3.6 Esterifications

Glycerol can be esterified with carboxylic acids or *via* carboxylation and nitration.[31] Reaction with carboxylic acids results in the formation of mono-acylglycerols and diacylglycerol. Monoacylglycerols are produced on a commercial scale by either continuous chemical glycerolysis of fats and oils (250 °C, alkaline, N_2 atmosphere) or by direct esterification with fatty acids.[32] The reaction of glycerol with dimethyl carbonate can also produce a high yield of glycerol carbonate in the presence of a biocatalyst (*e.g.* lipases). Glycerol can be converted to glycidyl nitrate by nitration that can be subsequently polymerised to form a valuable polymer.

1.3.7 Selective Oxidations

The oxidation of glycerol can be catalysed using highly active aerobic catalysts such as platinum and palladium. Supported gold catalysts are well known for catalytic stability, resistance to oxygen and tolerance against inhibition by aliphatic and aromatic amines. Organocatalysts such as 2,2,6,6-tetramethyl-piperidine-1-oxyl (TEMPO) can be used for the selective oxidation of glycerol to mesoxalic acid. TEMPO has also been used in electrochemical oxidation where glycerol is converted to 1,3 dihydroxyacetone (DHA). The reaction proceeds by applying a small electric potential to a solution containing glycerol, water and 15 mol% TEMPO using a glassy carbon anode. DHA can also be produced using biological oxidation *via* micro-organisms or enzymes. Other oxidation products include glyceraldehydes, glyceric acid, glycolic acid, hydroxypyruvic acid, oxalic acid and tartronic acid.

1.3.8 Pyrolysis

Glycerol was identified as a feedstock for pyrolysis in 1985, well before the growth in the biodiesel market. Recent research by Valliyapan and coworkers has focused on optimising conditions for hydrogen or syngas production.

Pyrolysis carried out in a continuous down-flow fixed-bed microreactor can take place with flow rates of nitrogen from 30–70 mL/min, temperatures of 650–800 °C and at atmospheric pressure. It was shown that the type and size of packing material in the tubular reactor can affect the conversion of glycerol and subsequent product distribution. Typical products include carbon monoxide, hydrogen, carbon dioxide, methane and ethane. At lower temperatures under steam or supercritical water conditions, longer molecules such as acrolein, formaldehyde and acetaldehyde are observed.

1.3.9   Biotransformations

Glycerol can be converted to a very large number of chemicals using micro-organisms and enzymes. The aerobic conversion of glycerol to 3-hydroxypropionaldehyde (3-HPA) was reported in 1985 by Slininger and Bothast.[36] The cells of *klebsiella pneumonia* can be grown on a rich glycerol medium and production of 3-HPA starts when these micro-organisms are added to a buffer containing semicarbazide and glycerol. It was shown that a yield of up to 84% could be obtained, however this yield is sensitive to cell age and cultivation medium. The optimal processing conditions for this experiment were 32 °C, pH 7–8 and glycerol concentrations of 20–50 g/L.

References

1. Joint Report Biorefinery Euroview (addenda D1.2 & D1.3) and BIOPOL (D4.2): Note with results identification, classification and mapping of existing EU biorefineries (www.biorefinery-euroview.eu).
2. The German Federal Government's Roadmap to Biorefineries (draft), published in 2011.
3. B. Kamm, P. R. Gruber & M. Kamm (ed.). 2006. *Biorefineries – Industrial Processes and Products. Status Quo and Future Directions. Vols. 1 and 2.* Wiley-VCH Verlag GmbH & Co. KGaA, Weinheim. 441 p. + 497 p.
4. F. E. I. Deswarte, J. H. Clark, A. J. Wilson, J. J. E. Hardy, R. Marriott, S. P. Chahal, C. Jackson, G. Heslop, M. Birkett, T. J. Bruce and G. Whiteley, *Biofuels, Bioprod. Bioref.*, 2007, **1**, 245–254.
5. S. Irmak and N. T. Dunford, Policosanol contents and compositions of wheat varieties, *J. Agric. Food Chem.*, 2005, **53**, 5583–5586.
6. V. Piironen, D. G. Lindsay, T. A. Miettinen, J. Toivo and A. M. Lampi, Plant sterols: biosynthesis, biological function and their importance to human nutrition, *J. Sci. Food Agric.*, 2000, **80**, 939–966.
7. Wellness Foods Europe Plant sterols, Opportunities for growth in the plant sterols market by Dr Franz Timmermann, Global Product Line Manager, Vegapure® plant sterols and sterol esters, Cognis Nutrition & Health, Germany.
8. D. M. Peterson, Oat antioxidants, *J. Cer. Sci.*, 2001, **33**, 115–129.
9. B. Matthaus, Antioxidant Activity of Extracts Obtained from Residues of Different Oilseeds, *J. Agric. Food Chem.*, 2002, **50**, 3444–3452.

10. Q. Chaudhry and D. Turley, www.sustoil.org.
11. B. M. Yapo, C. Robert, I. Etienne, B. Wathelet and M. Paquot, Effect of extraction conditions on the yield, purity and surface properties of sugarbeet pulp pectin extracts, *Food Chem.*, 2007, **100**, 1356–1364.
12. T. Bacic and Ming Long Liao, Sunflower Pectin: Adding value to agricultural biomass, CRC for bioproducts, University of Melbourne, www.australianoilseeds.com.
13. www.foodnavigator.com, WO2006/132288 A1.
14. B. Kamm, P. Gruber, M. Kamm, Biorefineries-Industrial processes and products, Wiley-VCH, Weinheim, 2006, pp. 295–315.
15. The Potential Industrial Uses of Forage Grasses Including Miscanthus P. A. Fowler, A. R. McLauchlin, L. M. Hall, *BioComposites Centre, University of Wales, Bangor,* 3 June 2003.
16. Ben Best, http://www.benbest.com/nutrceut/phytochemicals.html.
17. J. A. Hunt, *Pharm. J.*, 1999, **263**, 985.
18. M. Pagliaro, R. Ciriminna, H. Kimura, M. Rossi and C. D. Pina, *Angew. Chem. Int. Ed.*, 2007, **46**, 4434–4440.
19. G. Leffingwell and M. Lesser, *Merk index*, 11th edn., 1945, 705.
20. M. McCoy, *Chem. Eng. News*, 2006, **84**(6), 7.
21. J. Bonnardeaux, *Glycerin Overview*, Report for the Western Australia Department of Agriculture and Food, November, 2006.
22. K. D. Weiss, *Prog. Polym. Sci.*, 1991, **22**, 203–245.
23. M. Pagliaro and Michele Rossi, *The Future of Glycerol*, RSC, 2008.
24. A. Behr, J. Eilting, K. Irawadi, J. Leschinski and F. Lindner, *Green Chem.*, 2008, **10**, 13–30.
25. Y. Zheng, X. Chen and Y. Shen, *Chemical Reviews*, 2008, **108**, 5253–5277.
26. S. Yazdani and R. Gonzalez, *Current Opinion in Biotechnology*, 2007, **18**, 213–219.
27. G. Silva, M. Mack and J. Contiero, *Biotechnology Advances*, 2009, **27**, 30–39.
28. R. R. Soares, D. A. Simonetti and J. A. Dumesic, *Angew. Chem. Int. Ed.*, 2006, **45**, 3982.
29. K. Wang, M. C. Hawley and S. J. DeAthos, *Ind. Eng. Chem. Res.*, 2003, **42**, 2913.
30. H. Biebl, K. Menzel, A. P. Zeng and W. D. Deckwer, *Appl. Microbiol. Biotechnol.*, 1999, **52**, 289.
31. V. L. Budarin, J. H. Clark, R. Luque, D. Macquarrie, A. Koutinas and C. Webb, *Green Chem.*, 2007, **9**, 992–995.
32. N. O. V. Sonntag, *J. Am. Oil Chem. Soc*, 1992, **59**, 795.
33. M. J. Antal Jr., W. S. L. Mok, J. C. Roy and T. A. Raissi, *J. Anal. Appl. Pyrolysis*, 1985, **8**, 291.
34. T. Valliyappan, N. N. Bakhshi and A. K. Dalai, *Bioresource Technology*, 2008, **99**, 4476–4483.
35. W. Bühler, E. Dinjus, H. J. Ederer, A. Kruse and C. Mas, *J. Supercritical Fluids*, 2002, **22**, 37.
36. P. J. Slininger and R. J. Bothast, *Applied and Environmental Microbiology*, 1985, **50**, 1444–1450.

CHAPTER 2

Food Waste in the European Union

LUCY NATTRASS

NNFCC, Biocentre, York Science Park, Innovation Way, York,
YO10 5DG, UK
Email: L.Nattrass@NNFCC.co.uk

2.1 Introduction

Food waste includes food discarded during production, distribution, preparation and consumption. It comprises materials such as meat trimmings, bones and carcasses; fruit and vegetable trimmings, cores and rinds; leftover prepared meals, and out-of-date or spoilt ingredients. Food or parts of food items that were, at some point before disposal, edible are generally termed avoidable food waste, while materials that are not or have not been, under normal circumstances, edible are termed unavoidable food waste.[1] Some food waste is termed potentially avoidable, for example food material that some people eat that others do not, or materials that can be eaten when prepared in some ways but not others. The distinction between avoidable and unavoidable food waste is important because of differences in the causes of food waste, the potential to reduce food waste and the strategies employed to do this.

Food waste is generated at all stages of the supply chain, including in the production of fresh produce, during product manufacture and retail, in restaurants and other food-service operations, in the home, and by individuals away from the home, for example at work and outdoors.

Food waste is disposed of in separate food waste collections (source-segregated food waste), organic waste collections (mixed food and garden/green waste), as

RSC Green Chemistry No. 24
The Economic Utilisation of Food Co-Products
Edited by Abbas Kazmi and Peter Shuttleworth
© The Royal Society of Chemistry 2013
Published by the Royal Society of Chemistry, www.rsc.org

residual municipal waste, and as commercial and industrial waste. Liquid items may be disposed of to drains and in some places macerators are used to enable solid materials to be discharged to drains.

2.2 Causes of Food Waste

The majority of food wastage in the manufacturing sector is unavoidable.[2] This includes the inedible parts of animals, fruit and vegetables, discarded when produce is prepared for sale. Other sources of food waste in the manufacturing sector include operational errors and malfunctions that lead to previously edible food becoming waste.

Food waste is generated during distribution, storage and retail when it is not sold in time, or prematurely spoilt due to inadequate storage or damage. Fresh produce and products may be rejected and disposed of due to product specification. Product specification may relate to size, blemishes and aesthetic quality, and packaging, which may result in food products with packaging defects discarded when the quality of the food is not compromised.

Within households and the food-service sector, large quantities of avoidable food waste are generated. This includes food that becomes spoilt because it is not used in time, food that is prematurely spoilt due to inadequate storage, and food that is discarded due to consumer preferences or because excess food has been prepared.

2.3 Quantities of Food Waste Generated

The Swedish Institute for Food and Biotechnology (SIK) estimates that globally 1.3 billion tonnes of edible food waste are generated each year across food supply chains, including food losses during agricultural production, postharvesting handling and storage, and processing, and food wasted in distribution and consumption.[3] This is equivalent to around one-third of food produced for human consumption. On a per capita basis, much more food waste is generated in industrialised countries than in developing countries. In Europe and North America per capita food waste is estimated at 280–300 kg per year, whilst in sub-Saharan Africa and South and South East Asia it is 120–170 kg per year. However, food waste as a percentage of production is similar across all regions.

The best estimate of food-waste generation in the European Union is provided by the Bio Intelligence Service on behalf of the European Commission and is based on EUROSTAT waste statistics from 2006 and some national studies.[2] The study estimates that the European Union generates around 90 million tonnes of food waste per year, including food waste from manufacture, retail, and households, excluding waste generated in agriculture. This is equivalent to 179 kg of food waste per person per year. The same study predicts that, based on population growth and increasing disposable incomes, food waste in the EU could increase by around 40% by 2020 to 126 million tonnes per year, without additional prevention policy or activities.

2.4 EU Food Waste by Member State

Figure 2.1 illustrates that, of EU Member States, the UK generates the most food waste at over 14 million tonnes per year, followed by Germany, Netherlands, France, Poland and Italy. Per capita food waste is highest in The Netherlands at 579 kg per person per year, followed by Belgium and Cyprus at 399 and 334 kg per person per year, respectively, see Figure 2.2. Greece has the lowest per capita food waste generation at 44 kg per person per year. Some of these variations are due to the relative size of the food manufacturing industries within each Member State. The Netherlands, Belgium and Cyprus have the highest levels of manufacturing food waste per capita. In The Netherlands and Cyprus, manufacturing accounts for over 70% of total food waste, and over 50% in Belgium.

2.5 EU Food Waste by Sector

In the EU, the greatest quantities of food waste are generated by households, estimated at 38 million tonnes per year in 2006, equivalent to around 25% of food purchased, see Figure 2.3.[4] The manufacturing sector is the next largest contributor, producing around 35 million tonnes, while the food service and retail sectors generate smaller quantities of food waste at around 12 million tonnes and 4.4 million tonnes respectively. There are significant differences across Member States in the relative proportions of food waste generated in each stage of the supply chain, see Figure 2.4. For example, household food waste in the

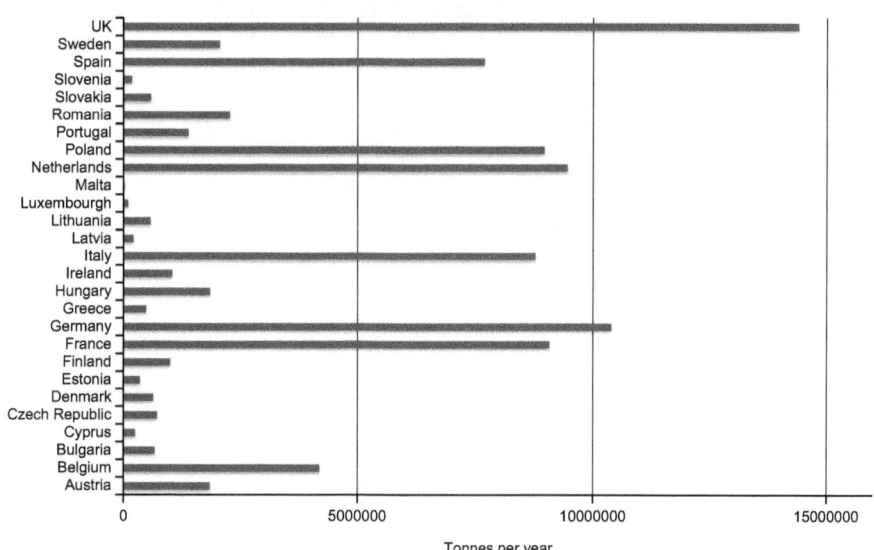

Figure 2.1 EU food waste by Member State.
Source: Bio Intelligence Service (ref. 2).

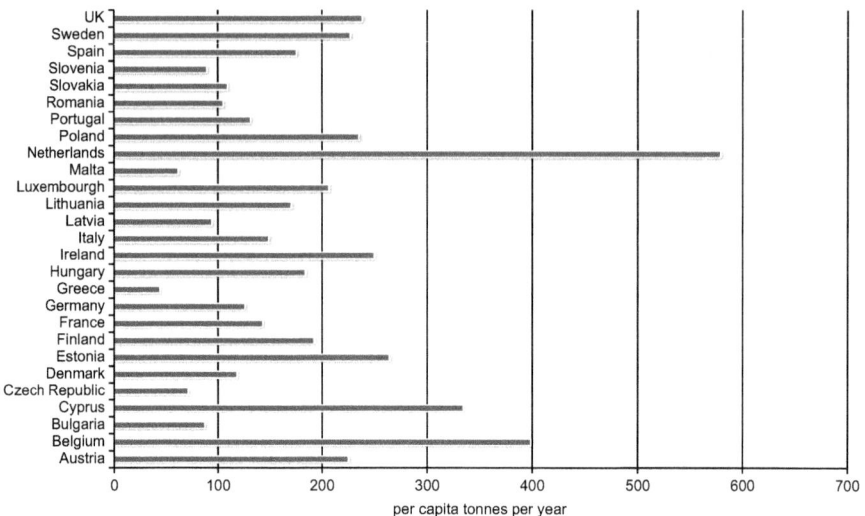

Figure 2.2   EU per capita food waste by Member State.
            Source: Bio Intelligence Service (ref. 2)

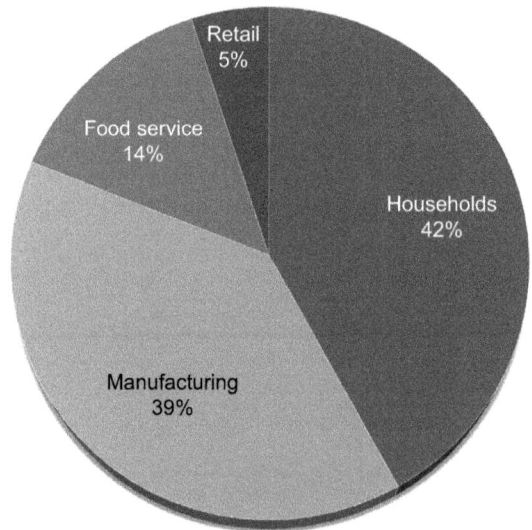

Figure 2.3   EU food waste by sector.
            Source: Bio Intelligence Service (ref. 2) and WRAP (ref. 4).

UK is around 137 kg per person per year; whilst in The Czech Republic and
Slovakia it is as low as 25 kg per person per year. In the EU15, food waste
generated in the food-service sector is estimated at 28 kg per capita per year,
whilst for the additional 12 Member States this is estimated at 12 kg/year.

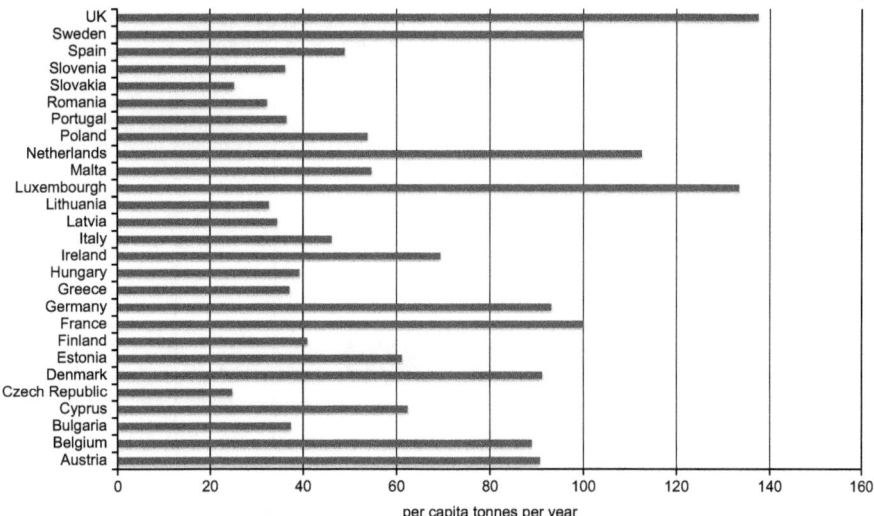

Figure 2.4 EU per capita household food waste by Member State.
Source: Bio Intelligence Service (ref. 2).

2.6 Food Waste by Food Group

The quantity of food waste generated is related to the type of food product; this may be due to the lifetime of perishable products, high susceptibility to damage, or high wastage related to low financial value. Figure 2.5 illustrates the estimated percentage of edible food wasted at each stage of the supply chain for each commodity group. These estimates are specific to Europe. Fruit and vegetables, including root vegetables typically generate more waste than meat and dairy products. For fruit and vegetables, including root vegetables, loss and wastage at agricultural production is greatest, whilst for cereals most wastage occurs at the consumption stage. In addition to edible food losses and waste, the food supply chain produces a large quantity of inedible or unavoidable waste, for example for wheat, barley and other cereals around 23% of the agricultural product is inedible, for fruit and vegetables the inedible fraction averages around 25%, and for fish and seafood the average is estimated at 50%.[3]

2.7 Data Sources

Under the European Union Waste Statistics Regulation (EC2150/2002) Member States must provide waste statistics to the European Commission every two years.[5] The statistics include an estimate of the quantity of waste generated, details on waste treatment and waste-treatment facilities. They are reported for the purpose of monitoring the implementation of waste policy.

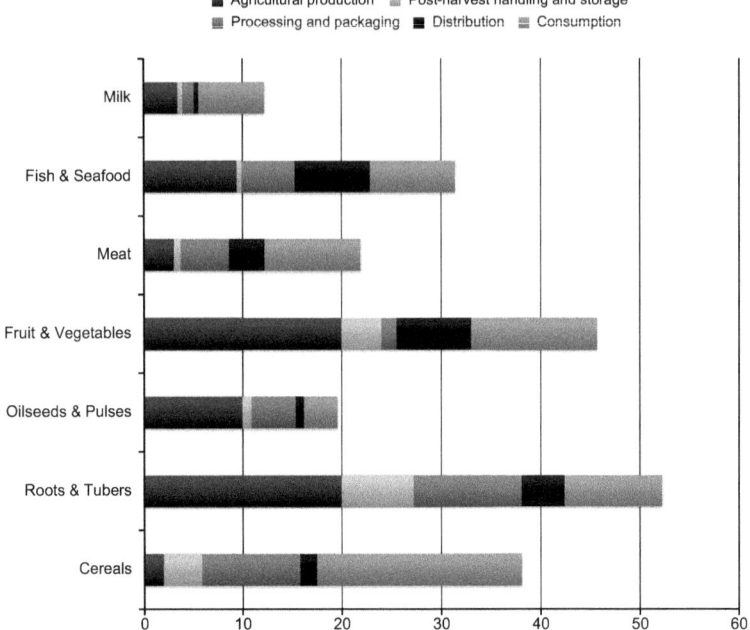

Figure 2.5   Estimated percentage of waste generate in each step of the food supply chain, FAO (ref. 3).

Waste quantities are reported according to a defined list of waste categories, and in relation to specified activities. For the identification of food waste the waste categories of interest are:[2]

- animal and vegetal wastes (excluding animal waste of food preparation and products; and excluding animal faeces, urine and manure); and
- animal waste of food preparation and products.

The activities of most interest in characterising food waste are:

- agriculture, hunting and forestry;
- fishing;
- manufacture of food products, beverages and tobacco;
- service activities, including wholesale and retail, hotels and restaurants, education, and health and social work; and
- waste generated by households.

At a Member State level waste data is collected by a combination of surveys, administrative sources, and statistical estimations. Administrative sources include data reported from licensed or permitted waste management facilities, and data provided by businesses under the Integrated Pollution Prevention and Control Directive. The Statistical Office of the European Communities

(EUROSTAT) compiles data from each Member State and publishes statistics at http://epp.eurostat.ec.europa.eu.

2.8 Data Gaps and Limitations

Food waste estimates are subject to uncertainty due to gaps in the available data. For example, in the waste statistics for 2006 five Member States provide no data on the quantities of waste generated by households. Large differences in household waste per capita may be due to inconsistencies between Member State reporting methodologies, which highlights a lack of standardisation in the allocation of data, and may result in green waste being included in some instances. Inadequate definitions contribute to the uncertainty of food-waste estimates, in particular the lack of clarity over the distinction between food waste and by-products. Tobacco production is reported with food manufacturing sector data, and the scope of service activities fails to provide a useful insight into the retail sector. The implementation and monitoring of future food waste policies may necessitate more accurate reporting.

2.9 Impact of Food Waste

2.9.1 Environmental Impacts

Food waste is predominantly biodegradable, and its disposal can therefore lead to the emission of greenhouse gases, including carbon dioxide and methane, to the atmosphere, while leaching can lead to the contamination of soils and water sources. Food waste also represents a waste of resources used in its production including land, water, energy, and fertilisers, and therefore life-cycle thinking or life-cycle analysis is the recommended approach to appreciate the full environmental impact of food waste. It is estimated that food waste generates at least 170 million tonnes of CO_2eq per year in the EU, equivalent to 3% of total emissions.[2]

The land resource wasted in the production of food waste should be appreciated in light of the increasing demand for land. Currently, food production makes use of 4.9 billion ha of land globally, equivalent to approximately one-third of land.[6] As the world population increases to over 9 billion by 2050, global food production must increase to meet the rising population, rising incomes and shifting diet. FAO estimate that global food production will need to increase by 70%, equivalent to between 1 and 2% per year.[7] The majority of increased production is expected to come from increases in yield and intensification of existing cultivated land, and around 20% from expansion of arable land. Water use in the agricultural sector is expected to increase from 2.7 trillion m^3 to between 10 and 13.5 trillion m^3 per year in 2050.[6] Reducing edible food waste could meet some of the additional demand for food, land and water.

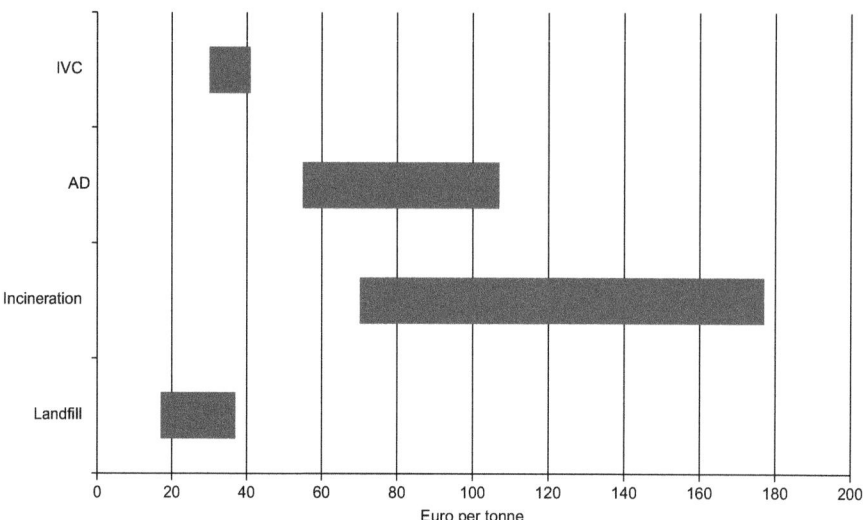

Figure 2.6 Cost of waste treatment options for food waste.
Source: ARCADIS and Eunomia (ref. 8).

2.9.2 Economic Impact

WRAP estimate that avoidable food waste represents on average £480 per UK household per year, equivalent to around 565 Euro[4] The actual cost of waste treatment ranges from 17–37 Euro per tonne for landfill to 70–177 Euro per tonne for incineration. The cost of invessel composting and anaerobic digestion are typically higher than landfill but less than incineration.[8] The costs illustrated here (see Figure 2.6) are calculated Net Present Values, at 2009, they exclude the cost of capital, the impact of incentives, subsidies and taxes and are not therefore indicative of gate fees that would include these items, and also reflect local conditions such as capacity, competition, and haulage.

2.10 Policy Framework

The generation, storage, transportation, management and disposal of food wastes are covered by various European and national policy instruments that relate to environmental protection, climate change, and resource management. These policies will impact the future quantities of food waste generated and how they may be used.

The European Commission launched an initiative for a resource-efficient Europe in 2011 under the Europe 2020 Strategy.[9] It aims to achieve sustainable economic growth by establishing an agreed long-term vision for resource efficiency and ensuring all related policies support this vision. The subsequent Roadmap to a Resource Efficient Europe provides a more detailed framework by which the EU economy could meet the 2050 vision in which all environmental assets are managed within their maximum sustainable yields, residual waste is reduced to almost zero, and ecosystems are restored. The Roadmap provides a framework that enables related policies to be implemented in a coherent and complementary way, and sets milestones by which to measure progress in improving resource efficiency.[10] The Roadmap includes milestones relating to transferring waste into a resource and increasing research and innovation for resource efficiency through supporting increased private sector investment. Food is highlighted as a key sector as the food and drink sector is responsible for 17% of the EU's direct GHG emissions and 28% of material resource use, the Roadmap targets a 50% reduction in edible food waste by 2020, and the Commission commit to further assess how best to limit waste throughout the food supply chain.

Relevant examples of existing EU measures include the "circular economy" strategy for the EU, building a recycling society with the aim of reducing waste and using waste as a resource, this is implemented by the revised Waste Framework Directive.[11] The revised Waste Framework Directive was adopted in 2008 and places an obligation on Member States to handle waste in a way that does not negatively impact human health and the environment, including the collection, transportation, treatment storage and disposal of waste. Under the Directive Member States have defined recycling targets up to 2020, and must ensure waste management practices comply with the waste hierarchy in which prevention is the favoured option, followed by reuse, recycling and recovery, with disposal as a last resort see Figure 2.7. Member States must prepare waste-management plans detailing waste quantities, sources and types, and the collection systems, and submit National Waste Prevention Plans by 2013, including food waste.

The policy framework places an impetus on Member States to facilitate the separate collection of biodegradable wastes. The Landfill Directive places an obligation on Member States to reduce the absolute quantities of biodegradable municipal waste disposed of in landfill.[12] By 2016, Member States must reduce the amount of biodegradable municipal waste sent to landfill

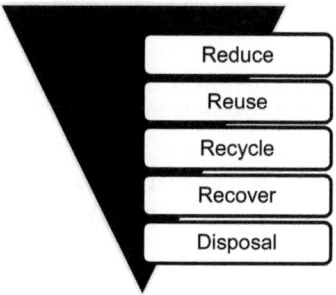

Figure 2.7 Waste hierarchy.

by 65% compared to 1995. The aim of the Landfill Directive is to reduce the GHG emissions from landfill, however, it does not prescribe alternative waste treatment options. In practice, Member States may opt for the least-cost waste treatment option and hence further regulation has been developed.

The revised Waste Framework Directive and the subsequent European Commission communications on biowaste management encourage the separate collection of biowastes, the treatment of biowaste in a way that fulfils a high level of environmental protection, and the use of environmentally safe materials produced from biowaste, with specific reference to composting and anaerobic digestion (AD). Biowaste is defined as biodegradable garden and park waste, food and kitchen waste from households, restaurants, caterers and retailers, and comparable waste from the food-processing sector.[13] Appropriately treated biowaste is counted towards Member States binding recycling targets for municipal waste. In addition, the European Commission committed to assess the appropriateness of a specific biowaste recycling target. The feasibility of biowaste recycling targets has been subject to consultation and the Commission has published two detailed impact assessments that detail policy options of implementing such a target and financial and environmental cost benefit analysis.[14,15]

End-of-waste quality standards are being developed so that compost and digestate products may be successfully marketed, and to enhance user confidence in the quality and safety of these products. The revised Waste Framework Directive also establishes a minimum efficiency threshold by which waste incineration plants may be regarded as recovery; below this threshold incineration is regarded as disposal within the waste hierarchy.

2.11 Waste Management

Current waste management practices for food waste include:

- landfill;
- incineration;
- animal feed;
- composting;
- anaerobic digestion.

There is no universal agreement on the most appropriate waste treatment option for food waste, as this depends on a number of local factors including the composition and quality of food waste, the collection infrastructure, and the demand for products such as energy and soil improvers (see Figure 2.8). The approach of each Member State typically can be described by one of the following three scenarios:[13]

- high levels of incineration employed to divert waste from landfill, high recovery including biological treatment;

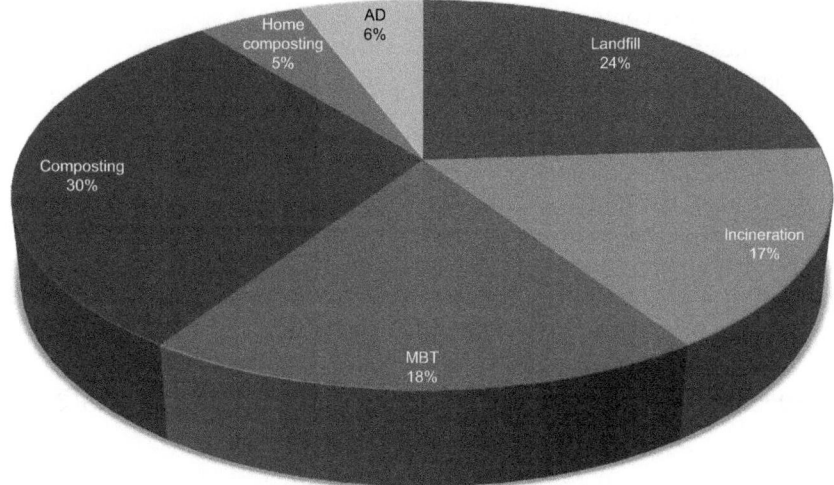

Figure 2.8 Waste treatment routes for municipal biowastes in the EU in 2013. *Source*: Bio Intelligence Service (ref. 15).

- high material recovery, predominantly organic recovery with lower rates of incineration; or
- landfill is the dominant waste treatment method, due to a lack of capacity for material recovery options.

2.12 Food Waste as a Biorefinery Feedstock

Growing interest in resource efficiency has highlighted the importance of reducing food waste and of diverting food waste from landfill into more valuable markets.

Drivers to using food waste:

- growing demand for renewable energy and fuels;
- need to divert biodegradable waste from landfill;
- interest in alternative feedstocks for chemical and material production.

A biorefinery involves an integrated series of complementary processes to produce a range of fuels, chemicals and/or materials from sustainable and renewable resources. Food waste could represent a low-cost and sustainable feedstock for biorefinery operations. However, there are a number of challenges to utilising food waste as a feedstock:[16]

- household food waste is dispersed, multiple collections are needed;
- heterogeneous physical and chemical composition, which changes due to seasonal variations;

- high moisture content;
- low calorific value;
- prone to bacterial contamination.

2.13   Reducing Food Waste

Under the waste hierarchy and resource efficiency agenda the need to reduce total volumes of food waste has received much attention. The degree to which strategies have been implemented to achieve this ambition varies between Member States. A number of UK initiatives aimed at various stages of the food supply chain are outlined in the case study below.

Case Study: UK Initiatives to Reduce Food Waste across the Supply Chain

- Courtauld Commitment: a voluntary agreement between food retailers, brand owners, manufacturers, suppliers and WRAP. It initially launched in 2005 with an aim to reduce household annual food waste by 155,000 tonnes between 2008 and 2010.

  Phase 2 of the Courtauld Commitment aims to reduce household food and drink waste by 4%, to reduce the environmental impact of packaging, and to reduce packaging waste. The activities undertaken by Courtauld Commitment signatories to achieve these goals have targeted products, packaging, shopping behaviours and the way the food is sold, and the behaviour of consumers in the home.[17]
- Love Food Hate Waste: a customer facing programme launched by WRAP in 2007, the programme aims to help households recognise and reduce food waste. The campaign promotes the benefits of saving money and reducing the environmental impacts of food waste, and specifically encourages consumers to take the following actions:[18]
  1. Check the contents of cupboards, fridges and freezers before shopping.
  2. Plan meals in advance and prepare shopping lists.
  3. Store fruit and vegetables in a fridge to extend the shelf-life.
  4. To ensure fresh food are stored in airtight containers to extend the shelf-life.
  5. Check use-by dates regularly and use products with the shortest shelf-life.
  6. Freeze food you will not eat before use-by dates expire.
  7. Measure portion sizes to avoid preparing and serving excess food.
  8. Use leftovers.
- Hospitality and Food Service Agreement: launched in 2012 by WRAP on behalf of UK, Scotland, Wales and Northern Ireland governments. The voluntary agreement aims to reduce food and associated packaging waste by 5%, and to increase the rate of food and packaging recycling, including AD and composting, to 70%.

References

1. WRAP. Waste Arisings in the Supply of Food and Drink to Households in the UK. 2010.
2. Bio Intelligence Service. *Preparatory Study on Food Waste Across EU27.* s.l. : European Commission DG Environment, 2010.
3. J. Gustavsson, C. Cederberg, U. Sonesson. Global Food Losses and Food Waste. Rome : FAO, 2011.
4. WRAP. Household Food and Drink Waste in the UK, 2009.
5. *Regulation (EC 2150/2002) of the European Parlimant and of the Councill on Waste Statistics.* s.l. : Official Journal of the European Communities, 2002.
6. Institution of Mechanical Engineers. Global Food: Waste Not, Want Not. London : s.n., 2013.
7. FAO. State of the Worlds Land and Water Resources for Food and Agriculture. Rome : s.n., 2011.
8. ARCADIS and Eunomia. Assessment of the Options to Improve the Management of Bio-wastes in the European Union. Brussels : European Commission DG Environment, 2010.
9. European Commission. COM(2011) 21: A resource-efficient Europe – Flagship initiative under the Europe 2020 Strategy. Brussels : s.n., 2011.
10. European Commissions. COM(2011) 517 final: Roadmap to a Resource Efficient Europe. Brussels : s.n., 2011.
11. European Parliment. DIRECTIVE 2008/98/EC on waste and repealing certain directives. *Official Journal of the European Union.* Brussles : s.n., 2008.
12. Council of the European Union. COUNCIL DIRECTIVE 1999/31/EC on the landfill of waste. *Official Journal of the EUropean Communities.* Brussels : s.n., 1999.
13. European Commission. COM(2008) 811: Green paper on the management of bio-waste in the European Union. Brussels : s.n., 2008.
14. COM(2010) 235: on future steps in bio-waste management in the European Union. Brussels : s.n., 2010.
15. Bio Intelligence Service. Assessment of feasibility of setting bio-waste recycling targets in EU, including subsidiarity aspects. Brussels : European Commission DG Environment, 2011.
16. C. S. K. Lin, L. A. Pfaltzgraff, L. Herrero-Davila, E. B. Mubofu, A. Abderrahim, J. Clark, A. Koutinas, N. Kopsahelis, K. Stamateatou, F. Dickson, A. Thankappan, Z. Mohamed, R. Brocklesby and R. Luque. Food waste as a valuable resource for the production of chemicals, materials and fuels. Current situation and global perspective. *Energy Environ. Sci.* 201, 2013, **6**, 426–464.
17. WRAP. Technical Memo: Evaluation of Courtauld Food Waste Target – Pase 1. 2010.
18. Love Food Hate Waste: An Introduction. s.l. : WRAP, 2010.

CHAPTER 3

The Thermochemical Conversion of Biomass into High-Value Products: Microwave Pyrolysis

PETER SHUTTLEWORTH,*[a] VITALY BUDARIN[b] AND MARK GRONNOW[b]

[a] Departamento de Física de Polímeros, Elastómeros y Aplicaciones Energéticas, Instituto de Ciencia y Tecnología de Polímeros, CSIC, Calle Juan de la Cierva 3, 28006 Madrid, Spain; [b] Green Chemistry Centre of Excellence, Chemistry Department, University of York, Heslington, York, United Kingdom, YO105DD
*Email: peter@ictp.csic.es

3.1 Introduction to the Biorefinery Concept

In order to achieve sustainable development mankind needs to move away from petrochemical feedstocks to renewable alternatives. A potential pathway to the production of alternative renewable products is the use of agricultural, forestry and municipal waste materials for the production of replacement products.[1]

Fuels derived from biomass are already established around the world using starch or sugar-based starting materials.[2] This method of biofuel production has received much negative attention due to competition with food production, use of valuable fertile agricultural land and intensive resource demands. Lignocellulosic biomass has the potential to provide up to 30% of US transport fuel demand without interfering with food supplies. The use of lignocellulosic

RSC Green Chemistry No. 24
The Economic Utilisation of Food Co-Products
Edited by Abbas Kazmi and Peter Shuttleworth
© The Royal Society of Chemistry 2013
Published by the Royal Society of Chemistry, www.rsc.org

biomass also offers the potential to convert agricultural and municipal waste into sources of energy.[3] As a solid fuel, biomass can be used in dedicated boilers or cofired with coal in existing power plants. Direct use is problematic due to the inherently high oxygen and moisture content, which result in a lower heating value. This, along with impurities such as chlorine and inorganic components, has led to problems with uptake of this technology.[4]

To overcome this problem a new biorefinery approach was developed to produce a high-quality drier fuel or energy-densified char that is more suitable for uptake in the energy sector. Biorefineries are facilities that integrate conversion processes and equipments to produce fuels, power, and chemicals from biomass.

At the moment there are two major approaches to convert lignocellulosic biomass to liquid biofuels: biochemical and thermochemical breakdown (see Figure 3.1), each of which comes with its own drawbacks. Biochemical conversion, through the use of enzymes, tends to be slow, requiring a batch-wise manufacturing process and can only be used to convert cellulose itself.[5] On the other hand, the process is selective and operates at low temperatures.

Thermochemical treatments tend to be relatively quick and continuous, involving high temperatures. High temperature initiates a number different processes resulting in a complex mixture of products with uncontrollable properties. For practical usage of biofuels they have to meet certain, controllable criteria. Char has to be grindable, hydrophobic and with as much energy density (calorific value) as possible. While bio-oils can be seen as a

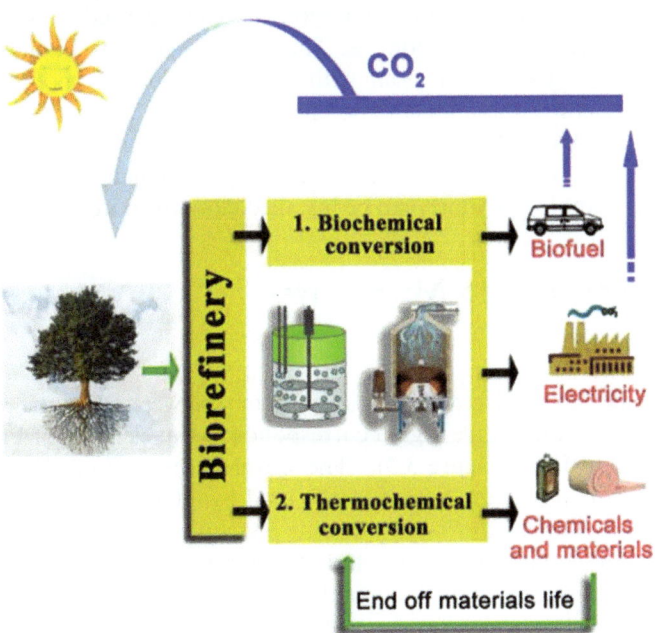

Figure 3.1 Types of biorefinery.

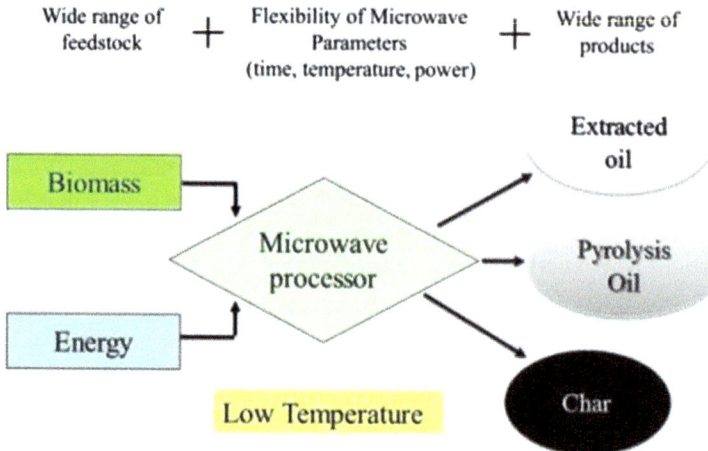

Figure 3.2   Principles of microwave biorefinery.

source of chemicals, they have also successfully been tested in stationary engines and boilers, and have been shown to be suitable for upgrading into high-quality hydrocarbon fuels for road and aviation uses.[6] An ideal conversion process would combine the advantages of both biochemical and thermochemical treatments, operating continuously, at low temperatures and producing a narrow product range on a reasonable timescale.

Here, we show that energy-efficient, low-temperature specific microwave activation of key plant structural components can be used to substantially enhance the energy value of biomass. The process converts biomass into an energy concentrated solid fuel, along with oils with properties superior to those achievable with conventional methods. This research demonstrates and delivers enormous economic and environmental benefits associated with the utilisation of biomass for a variety of higher-value energy products with flexible and controllable technologies which can be installed close to source (see Figure 3.2).

3.2   Introduction to Microwaves

3.2.1   Background

Microwaves lie in the electromagnetic spectrum between infrared waves and radio waves. This microwave region corresponds to wavelengths of 1 cm to 1 m (30 GHz to 300 MHz) (Figure 3.3).[7] The wavelengths 12.2 cm (2.45 GHz) or 33.3 cm (900 MHz) only are allowed to be used by international agreement for dielectric heating (unless thorough shielding precautions are taken) as the other frequencies are used for radar and telecommunications. The majority of domestic and commercially available microwave applications operate at 2.45 GHz as this frequency has the right penetration depth for heating food and also chemical reactions. In addition, the energy in a microwave photon (ca. 1 kJ mol^{-1}) is very low, relative to the typical energy required to break not

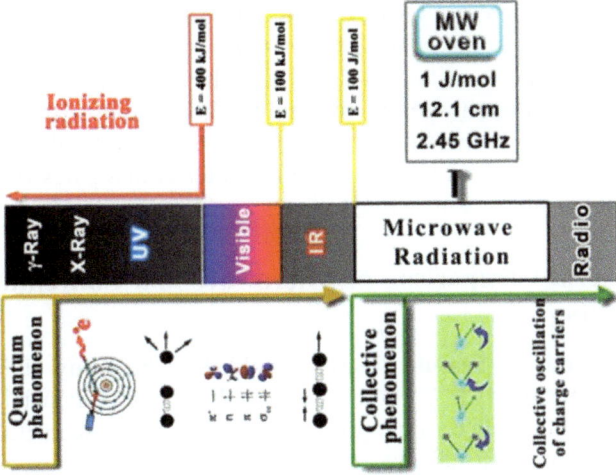

Figure 3.3 Microwave irradiation as part of the electromagnetic spectrum.

only a covalent bond (300–500 kJ mol^{-1})[8] but hydrogen bonds in water (ca. 20 kJ/mol).[9] Therefore, microwave excitation of molecules does not affect the structure of an organic molecule, and the interaction is due to purely the kinetic energy of the collisions.[10]

The microwave electromagnetic spectrum is divided into sub-bands, with the lower microwave frequency ranges (L bands) being used for the purpose of communication, and higher-frequency ranges (W bands) for analytical techniques such as microwave spectroscopy. In microwave spectroscopy photons of particular energies excite the rotation levels of gas-phase molecules. Whilst the absorption of microwaves in solid and liquid samples is frequency dependent; it is not quantised and therefore does not depend upon the absorption of a photon of the correct frequency. Rather, the material behaves as though it is reacting to a high-frequency electric field.[11]

3.2.2 Heating Effect

Microwaves are high-frequency oscillating electric and magnetic fields that affect anything that can be electrically or magnetically polarised. The heating effect arises from the interaction of the electric-field component of the wave with the material. There are three major mechanisms by which microwave radiation "heats" material: i) Conduction mechanism, ii) Interfacial polarisation and iii) Dipolar polarisation. The most common mechanism of interaction with dielectric materials, particularly with biomass, is the dipolar polarisation method. A material is classified as "dielectric" if it has the ability to store energy when an external electric filed is applied. The ability of the material to store electromagnetic energy within its structure is relative to the dielectric constant ε', however, in the presence of a varying electromagnetic field permanent dipoles inside dielectric materials are unable to follow the rapid reversals in the field. As a result of this phase lag, power dissipates in the materials. It is interesting to note that unlike conventional heating systems microwaves do not transfer energy through conduction or convection, microwaves transfer energy to the material by dielectric loss (power dissipated in a dielectric as the result of the friction produced by molecular motion when an alternating electric field is applied). The material dissipates this energy by dipolar rotation, and/or ionic conduction.

3.2.3   General Comparison of Microwave and Conventional Heating

Presently, microwave heating is well established in many industrial and commercial applications (*e.g.* food industry, pharmacy, polymer industry). It was shown that full-scale ceramic products could be processed with microwave energy faster, more cost effectively, and with equal or superior performance compared with congenital gas heating.[12]

Microwave irradiation is widely used in rubber vulcanisation. The comparative results obtained by applying separate electron beam irradiation and simultaneous electron beam and microwave irradiation show that the irradiation doses as well as irradiation times were markedly diminished, from 2 to 6 times, by simultaneous electron beam and microwave irradiation.[13]

It has been shown that microwave-assisted organic chemical reactions can be considerably more energy efficient than those using conventional heating and take significantly less time and require lower temperatures. A large number of examples highlighting the efficiency of microwave mediated reactions have been described in the literature, in particular in the areas of organic synthesis,[14] polymers,[15] and green chemistry.[16] It has also been demonstrated that microwave irradiation is a clean, cheap and convenient method in carbohydrate chemistry. Short reaction times and large rate enhancements as compared to conventional methods are observed; yields are comparable to or better than using conventional methods.[17]

These results show that the effect of microwave irradiation in chemical reactions is a combination of thermal heating, special thermal effects (localised heating) and nonthermal effects (highly polarising field).

In general, microwave heating for certain applications is a more efficient process than conventional heating and should be considered as an alternative and potentially faster, greener methodology (see Table 3.1).

It has been observed in many cases that thermal and nonthermal microwave effects could significant reduce temperature and time of chemical process and increase yield of product by comparison with conventional heating. The thermal effect (localised, selective heating) arises as a consequence of the inhomogeneity of the applied field, resulting in the temperature in certain zones within the sample being much greater than the macroscopic temperature. This overheating effect has been demonstrated in the decomposition of H_2S over c-Al_2O_3 and MoS_2-c-Al_2O_3.[21] Nonthermal effects are still a controversial matter: recently, Loupy and coworkers explained these effects as result from

Table 3.1 Comparisons of microwave and conventional heating technology.

Heating technology		
Conventional	*Microwave*	*Comments*
Slow and Superficial	Rapid and Volumetric	Microwave irradiation is rapid and volumetric, with the whole material heated simultaneously. In contrast, conventional heating is slow and is introduced into the sample from the surface. This feature is very important for processing of poor thermal conductors such as rubber and wood.[12]
Slow control	Instant control	Microwave heating can be controlled instantly and the power applied can be accurately regulated. This allows safe and precise control, even when applying very rapid heating rates.[8,18]
Nonselective	Selective heating	As with radio-frequency heating the power will selectively concentrate on the material that has the highest dielectric loss factor.[18]
–	Energy efficient	Calculations show that microwave-assisted organic chemical reactions can be considerably more energy efficient than those using conventional heating.[19]
–	Mobile	The advantage of microwave technology in terms of mobility and waste reduction applications have been highlighted by Ruan *et al.*[20]

modification of each of the terms of the Arrhenius equation (pre-exponential factor and activation energy) in the presence of a microwave field.[22,23]

3.3 Specialised Microwaves Techniques

3.3.1 Microwave-Assisted Heterogeneous Gas-Phase Catalysis

The application of microwave dielectric heating in a range of environment-related heterogeneous catalytic reaction systems has been reviewed by Zhang and Hayward.[24] The reactions investigated include decomposition of hydrogen sulfide, reduction of sulfur dioxide with methane, reformation of methane by carbon dioxide, and hydrodesulfurisation of thiophene. For example, decomposition of hydrogen sulfide into hydrogen and sulfur is commercially important for the coal and petrochemical industry.

Since the early 1980s, research laboratories have shown an increasing interest in microwave radiation as an energy source for the activation of oxidative coupling of methane in gas-phase catalysis. For example, oxidative dehydration of methane for synthesis of higher saturated and unsaturated hydrocarbons is a deeply investigated research field with the goal to access natural-gas deposits as a resource for the chemical industry in the context of progressively decreasing oil reserves. Under classical conditions, selectivity of 80–85% to C_2-hydrocarbons at methane conversions of 10–15% are common.[25] It was found that

Table 3.2 Influence of the heating method on the partial oxidation of methane.

Catalyst	T [K]	Conversion CH_4 [%]	Selectivity [%]		Heating method
			CO	CO_2	
10% Co/ZrO$_2$	723	48	79	21	microwave
	1073	100	99	1	microwave
	873	63	66	34	conventional
	1073	94	93	7	conventional
10% Ni/La$_2$O$_3$	673	58	74	26	microwave
	973	100	100	<1	microwave
	873	40	77	23	conventional
	1073	91	92	8	conventional

for $SmLiO_2$ and $(SmLiO_2)_{0.8}(CaOMgO)_{0.2}$ catalysts significant differences exist in the microwave field for conversions up to approximately 30 to 40%. While selectivity of C_2 approach zero under classical conditions, it almost reaches 100% in the microwave field. Recently, Bi *et al.* investigated partial oxidation of methane on mixed oxide catalysts.[26] The authors explain the temperature difference of 50–250 K for comparable methane conversions as resulting from the formation of hot spots on oxygen defects. A comparative overview of the experimental results is summarised in Table 3.2.

The selectivity in microwave-assisted heterogeneous gas-phase catalysis could be explained by the unique influence of microwave energy on the sorption process. Microwave energy may selectively heat the adsorbent surface and/or the adsorbed phase of the adsorbate, thus causing the gas phase and bulk solid phase to be at a lower temperature than that required for desorption by conventional heating. The temperature at the surface where sorption occurs is "effectively" greater than the measured solid or gas temperature; since oxides have a low permittivity and are relatively transparent to microwaves. For example, in the case of methanol and cyclohexane, microwave heating caused the methanol to desorb almost twice as much as conventional heating. This was because the methanol had a much greater permittivity (the bulk liquid permittivities have a 275:1 ratio).[27]

3.3.2   Microwave Plasma

Conventional gas-phase plasmas are formed by accelerating randomly occurring free electrons in an electric field until sufficient energy is built up to cause some of the gas molecules to ionise. Those electrons formed in the ionisation process subsequently accelerate further ionisation. This process results in the breakdown of the gas and formation of a plasma discharge. Plasma chemistry is a well-established field that focuses on the reactions of species found within the plasma and has found application in areas including substrate oxidation and anodisation, chemical vapour deposition and the Dupont arc acetylene process for conversion of methane to acetylene.[28] More recently, the use of microwave-induced plasma has been investigated for the cracking of methane to produce hydrocarbon fuels such as acetylene and has been identified as a possible selective and energy-efficient route for production of high-energy hydrocarbon fuels by cracking of low-energy hydrocarbons.[29–31]

Analysis of the gas composition of pyrolysed coffee hulls by both microwave and electrical heating suggested a different mechanistic decomposition route had taken place, with the possible formation of "microplasmas", which induce self-gasification of the char.[32] This hypothesis was corroborated by subjecting the char to reaction with CO_2 at different temperatures using both methods of heating. The results showed that, whereas the transition in the reaction mechanism controlling the Boudouard reaction (*i.e.* chemical or diffusion control) takes place at about 800 °C in conventional heating, in the case of microwave heating the temperature is much lower and the reaction never proceeds under pure chemical control, the differences between microwave and conventional heating being quite significant even at low temperatures.

3.3.3 Variable-Frequency Microwaves

Historically, there has been a lack of data on the dielectric properties of most materials as a function of temperature over the microwave range. The data that are beginning to emerge indicate that there is a complex interdependence of these properties on temperature and frequency.[12]

It is known in most cases that when low-loss materials exhibit an abrupt increase in the loss factor with increasing temperature,[33] the potential exists for very rapid changes in temperature. The temperature where this abrupt change in loss factor occurs is called the critical temperature. It is based on the difference in absorbed and dissipated power. However, some materials can display the opposite behaviour upon heating above a certain temperature, where the loss factor decreases significantly and heating becomes self-limiting. It has been shown that microwaves may offer self-limiting heating during polymer processing because materials are less susceptible to microwave heating once polymerised, and heating will focus automatically on the unreacted media.[12]

Metals, at least in bulk form, are excellent reflectors of microwave energy and in general are not heated significantly by microwaves. Other materials reflect and absorb heat to various degrees depending on their composition, structure, temperature, and the frequency of the microwaves. To illustrate this, pure water at room temperature is a broad-band absorber and absorbs well over a wide microwave frequency range, including 2.45 GHz. However, absorption is sensitive to frequency, and peak absorption for water at 20 °C occurs at about 18 GHz.[12]

Its absorption improves as ionic salts are added (*i.e.* composition change) due to increased ionic conduction losses. When water is heated, its absorption decreases at 2.45 GHz. Structural changes also affect its absorption. For example, when water is frozen, its absorption is negligible.

At room temperature, many ceramics and polymers do not absorb appreciably at 2.45 GHz. To efficiently heat a material that does not have an absorption mechanism close to 2.45 GHz requires changing either the frequency of the radiation or the composition of the material. Therefore, their absorption can be increased by increasing the temperature (in contrast to water), adding absorbing constituents (*e.g.* carbon, binders), altering their microstructure and defect structure, by changing their form (*e.g.* bulk *vs.* powder), or by changing the frequency of the incident radiation. Increasing the temperature (with radiant heat) is a common method used by many researchers to couple microwaves with poorly absorbing (low loss) materials. Once a material is heated to its critical temperature, microwave absorption becomes sufficient to cause self-heating. This hybrid method can result in more uniform temperature gradients because the microwaves heat volumetrically, and the external heat source minimises surface heat losses. Hybrid heating can be achieved either by using an independent heat source such as a gas or electric furnace in combination with microwaves, or through the use of an external susceptor that couples with the microwaves.[34] In the latter, the material is exposed simultaneously to radiant heat produced by the susceptor and to microwaves.

Another means of heating a noncoupling material is to incorporate absorbing additions. Reinforcements, binders, fillers, plasticisers, and dispersants are often added to polymers and ceramics as processing aids or to improve their properties. Proper selection of these additives can improve absorption and make them more favourable for microwave processing.[35]

As an example, Lambda Technologies, Inc. has commercialised variable-frequency microwave for materials processing. This type of microwave takes advantage of a broadband microwave sources to tune the incident frequency to the materials processed. Furthermore, the variable-frequency microwave system can sweep over several frequencies to achieve uniform distributions of electromagnetic energy and hence uniform heating of materials.[36]

Krieger has reported on the use of microwave vulcanisation, the first major application (other than food processing) of microwave energy in materials processing.[37] A key to its success was the strategy whereby the microwave system was designed as a vulcanisation processor and not as a microwave heating oven. Worldwide, there are over 600 lines producing thousands of tons of high-quality rubber.

In general, the areas of microwave processing can be divided into recovery (beneficiation of minerals and waste remediation); synthesis (pyrolysis, combustion, decomposition, catalysis); removal of phases (drying, solvent volatisation, binder burnout); fabrication (joining, CVI, coatings); consolidation (drying, sintering, melting, curing); and postfabrication treatments (annealing, surface modification, sealing, drying); and sample digestion for analyses.[12]

Sherritt, Inc., has used the microwave advantages of deep penetration and uniform heating to produce Si_3N_4 cutting tool inserts.[36] Scale-up of this process to a commercial level (over 500 inserts/24 h) resulted in improvements in both efficiency and product quality. The specific energy requirement was 3.6 kWh kg^{-1}, which is about a factor of ten less than the conventional electric-furnace process.

The effects of continuous and pulsed microwave irradiation on the poly-merisation rate and final properties have also been studied.[38] It was demonstrated that for certain epoxies, a pulsed microwave cure resulted in improved mechanical properties, better temperature uniformity, and a faster poly-merisation rate. Microwaves are also effective in the desulfurisation of coal and hydrothermal processing (see later).[39]

3.4 Selectivity of Microwave Technology in Organic Synthesis

There are numerous examples of selectivity enhancements using microwave irradiation in the literature, including transition metal, oxidation, reduction, heterogeneous catalysis and rearrangements.[40]

For example, selective oxidation of cyclohexene under microwave irradiation is reported whereby, depending on the reaction conditions, the epoxide (65% conversion, 75% selectivity), the enol (70% conversion, 80% selectivity) or the enone (>99% conversion, 89% selectivity) can be obtained, were a cobalt-salen-SBA-15 catalyst was employed.[41]

The orthogonal deprotection of alcohols can be achieved using neutral alumina under microwave irradiation, with chemoselectivity achieved, by simply varying the reaction time.[42]

Similarly, a simple method for oxidation of alcohols using clay-supported iron nitrate has been developed that selectively produces carbonyl compounds with no overoxidation to carboxylic acids.[43]

3.5 Comparison of Microwave and Conventional Pyrolysis of Biomass

3.5.1 Introduction

Pyrolysis is the thermal decomposition of organic matter in the absence of oxygen. The pyrolysis of biomass is a versatile process; relative yields of products can be controlled through variation of system parameters such as heating rate, maximum temperature and residence time.[44] Fast pyrolysis usually heats the biomass to the relatively moderate temperature of 500 °C, the short residence time, ~ 1 s, favours the production of a liquid portion ($\sim 75\%$) over gas ($\sim 13\%$) and char ($\sim 12\%$) on average. Increasing residence time to 10–20 s results in a significant change in the product distribution, $\sim 50\%$ liquid, $\sim 30\%$ gas and $\sim 20\%$ char on average.[45] It has also been shown that total liquid yield, increases as the quantity of lignin in biomass increase, while the ash and alkali metal content decreased.[46] The properties of the bio-oil liquid fraction depend strongly on factors such as biomass type, method of oil preparation, reaction conditions and the efficiency of sample collection.

The use of microwave pyrolysis of biomass dates to the early 1970s where Krieger and coworkers investigated the microwave treatment of wood, in addition to individual components such as cellulose and lignin.[47,48] The authors also investigated parameters such as temperature, pressure, time, power, particle size and density and looked at the effect of a number of these parameters on product distribution between char, gases and oils. They highlighted differences in the thermal behaviour of cellulose, lignin and hemicellulose under both conventional and microwave heating. It was found that the major products of cellulose decomposition were levoglucosan (27%), carbon dioxide (2–5%), water, and char. Total process times (even for large pellets) were reduced to less than 2–3 min when high-intensity microwave irradiation was employed. The gas composition in microwave pyrolysis of lignin differs from that of conventional pyrolysis studies due to the significant quantities of H_2 and C_2H_2 produced. This is in contrast to a high methane yield in conventional lignin pyrolysis, showing the selectivity advantage of microwave processing. Uniform thermal or volume heating occurs because the penetration depth of microwaves through lignin at this frequency is much greater than the pellet size.

Miura *et al.* looked at optimisation of microwave pyrolysis of cellulose and wood for production of levoglucosan.[49,50] They investigated particle size, power and irradiation time and source of biomass on product distribution in terms of water, oil, gas and char as well as conducting chemical analysis of the oil (main components: furfural, phenol, o-cresol, veratorol, guaiacol, levoglucosenone, xylosan, eugenol, galactosan, mannosan, and levoglucosan). It was demonstrated that temperature distribution, heat transfer and mass transfer are quite different from a conventional heating method. They also proved that microwave heating increased levoglucosan yield. Sarotti *et al.* have demonstrated that cellulose can be pyrolysed under microwave irradiation to produce levoglucosenone.[51]

Calorific values of microwave pyrolysis products of biomass have been investigated by Menendez and coworkers[32,44,52,53] Calorific values of the collected fractions were up to 7, 37 and 10 kJ/g for char, oil and gas, respectively, with relative weight amounts up to 10, 3 and 26% (with up to 69% water). The calorific value of obtained oil is relatively close to typical values for petroleum derived oil (42 kJ/g) although yield was relatively low. Menendez and coworkers have also investigated the effect of temperature on relative yields of different pyrolysis fractions and elemental composition of chars and oils and chemical composition of gas. The results show that oils from microwave pyrolysis are more aliphatic and oxygenated than those produced by conventional heating at the same temperature (1000 °C). They also found that microwave pyrolysis produces more gas and less oil than conventional pyrolysis and additionally the amount of hydrogen in this gas is much higher especially at lower temperatures. The same team also described the conventional and microwave-assisted pyrolysis of coffee hulls at 500, 800 and 1000 °C. The calorific value of the oils and gases are higher when MW pyrolysis is used compared to conventional pyrolysis, while char values are similar for both pyrolysis methods (24 kJ/g). The effect of temperature on the calorific value of the oils and chars is practically negligible. However, the calorific value of the gases increases with increase in pyrolysis temperature, the oil values varying from 30–32 kJ/g in conventional heating to 34 kJ/g in MW. For the gases, it was observed that at 500 °C the gas obtained from MW pyrolysis exhibited a calorific value twice that of gas from conventional pyrolysis.

It has been shown that MW pyrolysis of the coffee hulls produces more gases but fewer oils than conventional pyrolysis; the char yields being virtually identical for both pyrolysis methods. This suggests that in microwave treatment secondary cracking reactions of the oil components occur to a greater extent than in conventional heating.

Use of microwave irradiation in low-temperature pyrolysis is relatively unexploited. Ruan and coworkers have investigated microwave biomass pyrolysis at temperatures between 200–600 °C, using a variety of biomass sources including cellulose, pine, corn stover, canola, municipal waste and aspen.[20,54] They also investigated temperature optimisation to minimise side reactions and gain a greater control of the product composition. Analysis of pyrolysis oil in terms of water content, density, pH, viscosity, elemental analysis and calorific value were reported.[55] The properties reported, including calorific values, were in the range of similar bio-oils (19 kJ/g) produced by conventional pyrolysis process. The influence of pyrolysis conditions such as irradiation power, temperature and time on elemental and chemical composition of gas and chars were also presented. The authors reported that at elevated temperatures they observed increased yields of hydrogen gas and decreased yields of char.

More recently, microwave-induced pyrolysis at temperatures between 250 and 560 °C has been investigated by Huang et al.[21] They looked at pyrolysis of rice straw and identified solid, liquid and gaseous products. The microwave power and particle size of feedstock were both key parameters affecting the performance of microwave-induced pyrolysis. After the pyrolysis of rice straw, three main products were generated and collected separately. About half of rice straw sample was transformed into H_2-rich fuel gas (H_2, CO_2, CO, CH_4, respective percentages of 55, 17, 13, 10 vol.%).

3.5.2 Low-Temperature Microwave Processing

Microwave treatment of biomass has been focused on high-temperature pyrolysis, gasification and liquefaction of the starting materials. Pyrolysis of biomass is usually conducted at temperature above 350 °C. Examples of substrates employed in this type of pyrolysis are numerous and include plant biomass such as wood or agricultural residues, plastics and municipal waste. It has been shown that there are a number of advantages of microwave pyrolysis above conventional due to high rate of pyrolysis, better controllability, energy efficiency and selectivity. This selectivity may be in the ratio of solid–liquid–gas, or it may be in composition of the bio-oil.

3.5.2.1 Cellulose Pyrolysis

Lignocellulosic biomass consists of three structural polymers (hemicellulose, cellulose and lignin) along with various extraneous components.[56] These polymers differ greatly from each other in terms of functional groups, molecular weight, structural order and accessibility. Due to high level of crystallinity activation of cellulose is one of the biggest challenges of biorefinery. A number of efforts have been made to activate cellulose by microwave irradiation.

In order to differentiate thermal and nonthermal events during microwave treatment an experimental setup was designed based on a narrow packed column (allowing homogeneous exposure to the microwave irradiation) in a well-stirred oil bath within the microwave chamber (to ensure uniform heating throughout the sample) with the top portion of the cellulose protected from microwave irradiation using aluminium foil (Figure 3.4A). The experiment clearly illustrated the microwave effect on cellulose. The cellulose that had been exposed to microwave irradiation visibly began to decompose, while the area above shows little or no decomposition (Figure 3.4B). A range of experiments

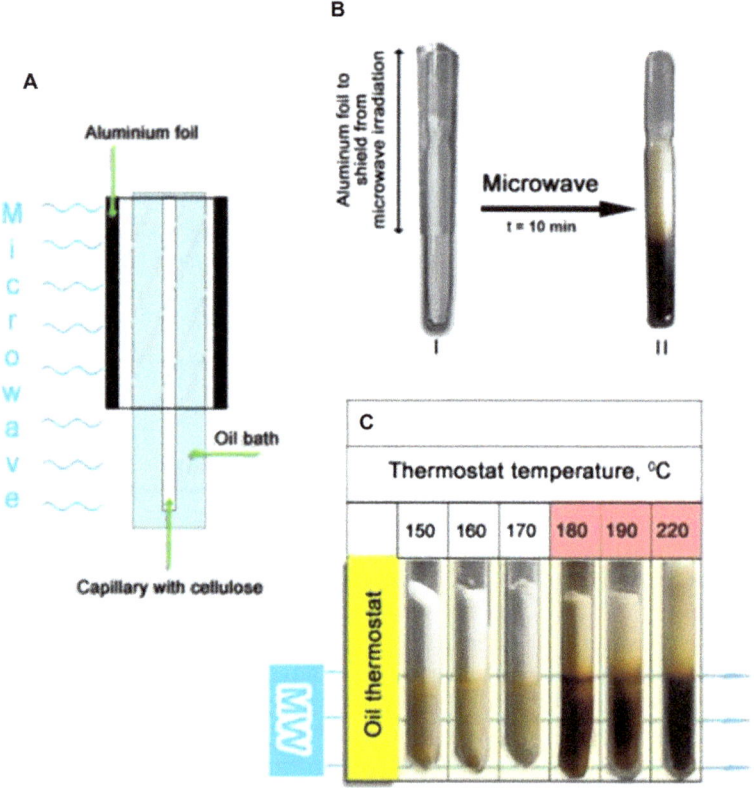

Figure 3.4   Scheme of experimental setup for estimation of direct microwave effect on the cellulose. (A) Construction of the cell. (B) Image of cellulose in capillary before (I) and after (II) microwave treatment. (C) Cellulose samples treated in the microwave from 150–220 °C under the sample processing condition as those in (B).

were carried out between 150–220 °C (Figure 3.4C). A marked change was seen to occur around 180 °C. Above this temperature the cellulose is seen to decompose to a greater extent producing a solid char along with an oily liquid fraction.

Since 180 °C had been established as a significant temperature point in terms of cellulose decomposition, thermal investigations of cellulose were carried out to find a connection between this temperature and any thermal events that may be taking place within the polymer at this temperature. A modulated differential scanning calorimeter (MDSC) has been chosen as a preferred method of characterisation. This technique simultaneously provides increased sensitivity over traditional DSC along with better resolution, due to a combination of low average heating rate and following a frequency-dependent modulation. It was found that within cellulose structure a phase transition takes place at approximately 185 °C, which is in a good correlation with specific temperature of microwave-assisted decomposition of cellulose. Above this temperature, the specific microwave effect becomes pronounced and the rate of degradation of cellulose increases significantly.

The combination of experimental techniques proves that the microwave degradation starts within the amorphous region. The enhanced molecular freedom within this region resulting from the phase transition allows improved interaction between the microwave energy and cellulose. Below 180 °C the polar groups in cellulose have less freedom so cannot rotate easily, resulting in a poorer interaction. Above 180 °C, as the number of groups capable of rotating increases particularly, but not exclusively, in the amorphous region the rate of decomposition increases.

It is known that the structure of cellulose within the cell wall of plants exists in a fibrillar form. Within these fibres the amorphous and crystalline regions alternate at a period of approximately 15 nm (Figure 3.5A). Crystalline cellulose contains a very ordered hydrogen-bonded network within which a proton-transport network is possible in the presence of an electromagnetic field (Figure 3.5B). Below 180 °C the hydrogen-bond network within the amorphous regions is relatively small and localised. For these reasons cellulose fibres could be represented as alternating ionic conducting (crystalline) and nonconducting (amorphous) regions. When the crystalline region is placed within an electromagnetic field it will polarise, generating a charge on the crystalline interface (overall charge will be zero). The presence of acid has been found to promote char formation. At temperature above 180 °C the acidity in the amorphous region is increased as the softening of the polymer allows movement of protons from the crystalline region, causing acid-catalysed decomposition of cellulose (Figures 3.5C and D). Analysis of liquid fractions produced during microwave pyrolysis of cellulose has shown significant levels of acids along with the furans and substituted phenols (Budarin *et al.*, 2009).[57] The production of these components will further plasticise the amorphous region, exacerbating the enhanced microwave effect. Increasing temperature and acidity will eventually disrupt the crystalline regions of cellulose, aiding its degradation to char and volatiles.

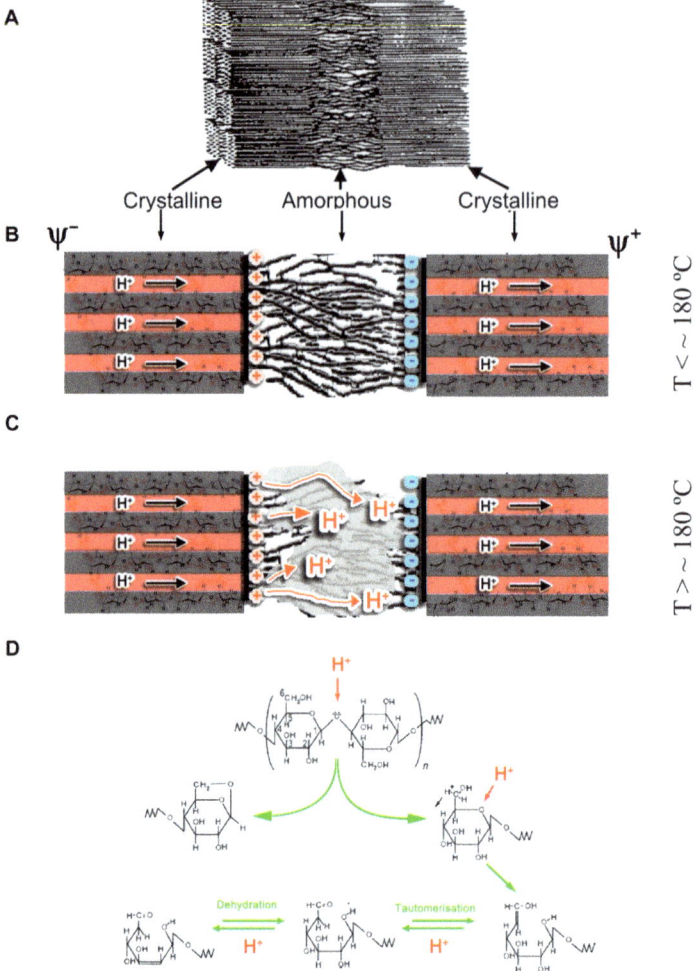

Figure 3.5   Mechanism of microwave interaction with cellulose.

3.5.2.2   *Macroalgae Microwave Pyrolysis*

It has been suggested that marine biomass could be a promising alternative to terrestrial grown biomass for fuel replacement.[58] This alternative that comprises up to 50% of the global biomass typically has high growth rates.[59] Seaweed, which has many commercial outlets in food, medicine and fertilisers, has one of the highest solar energy conversions, with currently less than 1% of it currently being utilised.[60] The macroalgae seaweed, unlike microalgae, is also easy to extract, the main problem being the means to utilise such vast quantities of this material (>10 million tonnes) effectively.[61]   A readily transportable microwave technology could alleviate this problem, transposing this under-utilised biomass into a future renewable source of chemicals and fuel.

Buderin *et al.* demonstrated the effectiveness of microwave-mediated pyrolysis of *Gracilaria* macro-algae. It was proved that this biomass source could be effectively broken down into useful chemicals at temperatures close to 130 °C, which is exceptionally low for such processes, and hence, potentially more energy efficient than other pyrolysis means.[62] In addition, it was found that it could be carried out in aqueous media, eliminating costly drying steps and that there was a critical mass point in which the distribution of products changed dramatically. At low biomass concentrations it was found that a slower pyrolysis rate occurred, resulting in a final product distribution of fermentable anhydro-sugars (*i.e.* levoglucosan). At high mass concentrations the rate of reaction dramatically increased, with the final product distribution being phenol and 4-methyl phenol. This leads to the exciting potential to tailor the process to product requirements, with the products formed in both cases being highly sought after.

3.5.2.3 Orange-Peel Valorisation Using Microwaves

Orange processing consists of up to 90% of the largest fruit crop, citrus fruit that is grown on a scale of over 110 million tonnes per year.[63,64] In addition, up to 60% of the fruit is discarded or pelletised for animal feed after juicing etc.[65] This waste orange peel has been shown to produce valuable products such as D-limonene, pectin and a form of mesoporous cellulose without the need for pretreatment using microwave hydrothermal processing (170–200 °C).[66,67] After this success, the experimental procedure was scaled to three litres that was found to increase the attainable yield of D-limonene from 0.19% (25 mL scale) to 1.52%.[68] The produced D-limonene was then used to produce a sustainably derived *p*-cymenesulfonic acid that had comparable acidic catalytic properties to that of the much used petroleum derived *p*-toluenesulfonic acid.[69]

3.5.2.4 General Biomass Microwave Treatment

From the data above and from Table 3.3 it can be seen that the yield of pyrolysis oil is dependent on initial conditions (*e.g.* nature of feedstock,

Table 3.3 Mass balance of what straw microwave pyrolysis.

				Yield (mass%)			
Feedstock	*MW Power*	*T/°C*	*additive*	*Char*	*Organics*	*Water*	*Gas*
Wheat straw	500	130	–	29.4	15.2	37.2	18.2
Wheat straw	1000	130	–	29	20.6	36.4	14
Wheat straw	1200	140	H_2SO_4	44.4	7.3	32.9	15.4
Wheat straw	1200	140	HCl	31.8	22.1	27.1	19
Wheat straw	1200	165	NH_3	40.7	17.0	22.3	20
Pine wood	1200	140	–	32.7	26.1	27.7	13.5

temperature, microwave power, additives). Increasing microwave temperature and power typically increase the yield of oil. Thus, increasing microwave power from 500 to 1000 W increases the yield of wheat straw bio oil by 1.3; addition of 3% sulfuric acid to wheat straw halves the yield of bio-oil. It should be noted that addition of hydrochloric acid to wheat straw causes an increase in bio-oil yield to 22%. We believe that optimisation of microwave parameters for specific feedstock's will significantly increase bio-oil yield.

A comparison of mass balance data for wheat straw pyrolysis form Table 3.4 (see below) and experimental data (Table 3.3) shows that in the presence of hydrochloric acid microwave pyrolysis produces very similar quantities of bio-oil to the conventional process. Since wheat straw is very high in potassium content, one would expect a significant increase in char yield (and not bio-oil), for which the microwave processing seems to compensate (Table 3.5).

Our data about temperature/microwave power influence on the mass balance of microwave pyrolysis are in good agreement with data from the literature.[47]

3.5.2.5 Verification of Sample Temperature Measurement during Microwave Experiment

The measurement of temperature is a key issue for both microwave chemistry and pyrolysis. Bulk temperature cannot be directly measured as inside the microwave cavity most temperature probes will be either directly heated at a different rate from the substrate under investigation and by the nature of temperature as a concept it cannot be measured in a nonequilibrium energy-transfer situation as occurs under microwave irradiation (*i.e.* direct excitation of rotational levels leads to non-Boltzmann distribution), but only after this energy has been dissipated (probably relatively quickly). External temperature measurement systems suffer from the traditional issues of lag time caused by the heating of the material followed by transfer to the vessel by conduction.

In essence, we can see from the discussion above, one should have limited confidence in temperature measured within a microwave cavity as a consequence of the physical chemistry occurring. Furthermore, it is well known that within microwave fields, due to differences in absorption of microwave energy it is possible to have hot spots with temperatures far in excess of the bulk temperature.[7]

Thorough work by Delft TU of the Netherlands investigated the issue of temperature measurement within microwaves, the associated accuracy or inaccuracy, probe application and spatial variation.[70] In their work, it was observed that there was a considerable but predictable difference between an insample fibre-optic probe and exterior infra-red detection. During microwave pyrolysis a difference in temperature measurement between these two probes also has been observed (see Figure 3.6).

However, below 100 °C the reproducibility between the two methods (Figure 3.6) is good; this could be due to water vapour being evenly distributed within the biomass bulk. At temperatures higher than the water boiling point a difference exists between the measurement methods, due to hot spots, however, as volatiles are evolved the hot spot is cooled by evaporation of pyrolysis

Table 3.4 Mass balance of reference fuels and Lolium/Festuca grasses.

Feedstock	Willow	Switch grass	Switch Grass W	RCG	Wheat Straw	Dactylis Glomerata	Festuca Arundin acea	Festuca Arundinacea W	Lolium perenne
Reactor temp./°C	507	500	509	502	509	500	510	504	503
Hot vapour residence time (s)	0.75	0.91	0.76	0.73	0.82	0.88	0.84	0.89	0.93
Moisture content	7.8	8.3	5.0	7.9	9.1	5.1	4.7	5.0	4.8
Ash (mf wt%)	1.34	4.3	3.4	5.1	6.3	7.5	7.3	4.4	6.2
Char	20.9	24.7	20.2	22	31.9	34.9	33.8	19.9	32.1
Organics	52.9	51.5	55.2	47.2	24.9	30.6	30.8	41.7	29.5
Gas	9.3	7.9	10.54	11.1	15.6	13.7	15.4	21.6	14.2
Water in feed	15.9	12.4	11.9	13.1	25.7	20.8	16.4	8.7	24.2
Total liquids	68.9	63.8	67.1	60.2	50.5	51.4	47.2	50.4	53.7
Closure (%)	99.0	96.5	97.8	93.4	98.2	93.5	96.5	91.9	92.5

Table 3.5 Mass balance comparison of conventional and microwave pyrolysis of wheat straw.[44]

| | Yields (mass%) | | |
Products	Conventional	MW 1000 W	MW, 1200W 3% HCl
Char	31.9	29	31.8
Organics	24.9	20.6	22.1
Water	25.7	36.4	27.1
Gas	15.6	14	19

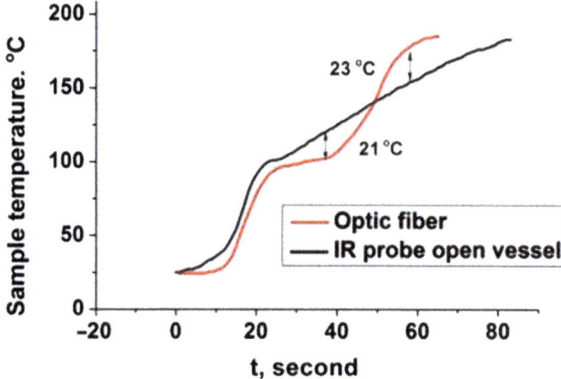

Figure 3.6 Temperature profile of barley dust in an open vessel microwave using differing testing methods.

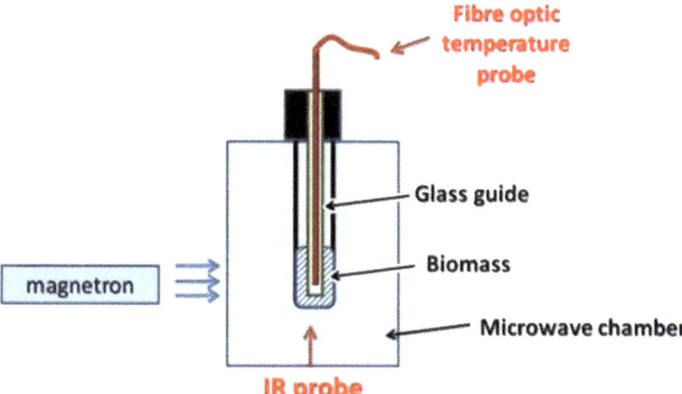

Figure 3.7 Temperature measuring devices within a microwave cavity.

product and these gaseous products distribute the heat more evenly through the sample. The fibre-optic measurements are alternating above and below the IR trend due to the response lag of the method. The fibre-optic probe is held within a glass guide tube that leads to a delay in heat transfer to the probe (Figure 3.7).

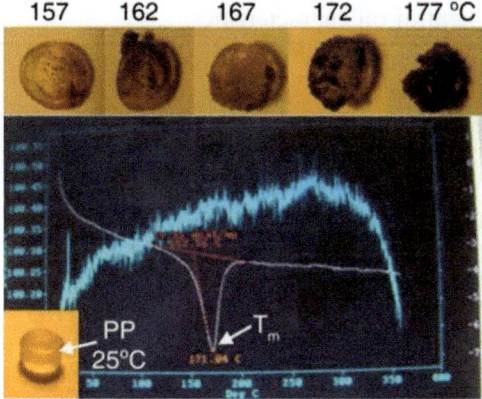

Figure 3.8 Polypropylene used as an in situ temperature verification experiment. Left, polypropylene bead during various stages of melting. Right, an STA trace showing the melting point of polypropylene.

To further add confidence to the hypothesis that temperature measurement by external temperature probes were correct, experiments were carried out using a novel technique of adding nonmicrowave adsorbing polymer beads of known melting point to the reaction mixture. Reactions are carried out at a range of temperatures – as observed by the external infra-red probe. The results are shown in Figure 3.8.

As can be seen in the photographs of the melted polypropylene bead the probe-measured temperature follows the melting trend of the polymer bead (as shown in the STA trace), which gives confidence in the infrared temperature measurement technique as providing a fair and reliable representation of the temperature of the bulk material during microwave pyrolysis of biomass.

3.5.3 Pyrolysis Oil Requirements – Chemical or Fuel Replacement

3.5.3.1 Fuel Replacement

High-grade pyrolysis oil is oil that has both physical and chemical properties that are dramatically improved in comparison to those of conventional biomass pyrolysis oil, and are closer to the properties of crude oil itself. It has been demonstrated that bio-oil produced via microwave pyrolysis has great potential, in that the water content and total acid number are exceptionally low for pyrolysis oil types.

Key parameters that need to be assessed in order to quantify the "grade" of oil are outlined in Table 3.6.

3.5.3.2 Chemical Source

Wheat straw is an agricultural residue high in the main plant nutrients, especially potassium that is a well-known catalytic metal for biomass pyrolysis.

Table 3.6   Assessment of oil characteristics.

Property	Crude oil[71]	Pyrolysis oil	Microwave oil		
			Menendez	Ruan[55]	York
Water (%)			–	15.2	<1
C (wt%)	85–87	45–55	75.34	60.1	58.9
H (wt%)	10–14	6–7	7.92	7.70	6.85
N (wt%)	0.1–2	0.3	5.64	2.02	1.15
O (wt%)	0.1–1.5	40–50	10.72	29.4	33.2
S (wt%)	0	0.5–5	0.63	0.15	0.02
Specific gravity	0.8–1.0	1.2–1.3	–	–	1.2
Acid number	<1	70–150	–	pH = 2.87	1.4
Alkali metal	50	100	–	7.6	5.71
LHV (kJ/g)	42	16–21	36.1	17.4	16–22

Note: Lower heating value (LHV) is related to calorific value.

The material shows, under fast pyrolysis (600 °C) conditions, that the decomposition of lignocellulose proceeds via an ionic mechanism.[72–74] Thus, low yields of sugars and high yields of furfural and acetic acid are observed from cellulose and hemicellulose decomposition.[75] A mixture of phenols and methoxyphenols are produced from decomposition of the lignin component of the biomass.[46,76]

Under microwave conditions the pyrolysis oil from wheat straw (control experiment) show catalysed decomposition product markers (furfural and acetic acid), however, the product distribution contains higher relative yields of phenols, methoxyphenols, diphenols and di-methoxyphenols, and high relative yields of the primary cellulose decomposition products (levoglucosan and 1,4:3,6-dianhydro-α-D-glucopyranose), possibly due to the milder temperatures experienced in this case. Also small relative yields of long-chain acids, such as hexadeanoic acid are present. Thus, qualitatively, it appears that the microwave pyrolysis oil has a higher relative yield of aromatics and primary sugars compared to the fast pyrolysis oil. In addition, the microwave pyrolysis oil shows a much-reduced quantity of acetic acid and "cleaner" product distribution. Furthermore, the microwave-processed bio-oil is one-third made up of two components, levoglucosan and 2,3-dihydrobenzofuran. Both levoglucosan and 2,3-dihydrobenzofuran can be isolated and used as feedstocks for chemical processing, hydro-deoxygenated to improve fuel characteristics, or indeed levoglucosan could act as a feedstock for bioethanol production.

In particular, processing in the presence of sulfuric acid causes the bio-oil to be nearly 50% levoglucosenone and levoglucosan combined. The high relative yields of levoglucosan suggested less cracking/rearrangement. The formed levoglucosenone is a known marker in acid-catalysed rearrangement/dehydration of cellulose during pyrolysis.[77,78]

The influence of a basic atmosphere (ammonia) during microwave pyrolysis was found to be similar to the control experiment, suggesting it has little influence.

Table 3.7 Bio-oil upgrading routes.

Upgrading	Target	Needed	Effects
Esterification	Acids	Catalyst + alcohol	Reduced corrosiveness Improved stability Increased energy content
Hydrogenation	Alkenes aromatics	Catalyst + hydrogen	Increased energy content Improved stability
Reduction	Aldehydes	Catalyst + hydrogen	Improved stability Potential for using alcohol products to esterify acids

3.5.4 Pyrolysis Oil Upgrading

The negative properties of bio-oil such as high viscosity, poor storage stability and corrosiveness stand in the way of the large-scale uptake of bio-oil as a conventional oil substitute. As a result, much research has been carried out into processes that will make it more suitable for the desired application. The main upgrading routes listed in the literature include hydrogenation, esterification, hydrodeoxygenation, catalytic cracking of the pyrolysis vapours, steam reforming and the physical emulsification of bio-oil with diesel. Although hydrogenation and esterification are less atom-efficient processes they are much simpler than the others, all of which require complicated expensive procedures and rely on the use of catalysts that are subject to fouling during processing.[79–81]

General upgrading schemes for pyrolysis oils include (also see Table 3.7):

- esterification of acids to reduce acidity increase stability and to add alkyl groups in order to increase energy content (residual alcohols can also improve homogeneity and flow characteristics);
- hydrogenation of unsaturated groups to provide additional hydrogen and therefore energy;[82]
- reduction of aldehydes to improve stability and burning characteristics.

In addition, the simple addition of small alcohols has been shown to improve stability and fuel performance.[55]

3.6 Summary

Microwave heating offers a uniform, rapid, instantaneous, controllable and tuneable method for heating materials and thereby initiating chemical conversions. From the wide range of examples the almost universal appeal of this method can be observed. The major flaws of microwave heating have been turned to advantages. Limited penetration means small continuous reactors can be employed, thereby minimising the inventory of active material means a safer system. The relatively high energy cost is offset by cleaner products in higher yields and in some cases products that could not otherwise be observed.

Acknowledgements

The authors would like to thank past and present members of the Green Chemistry Centre of Excellence for their input and useful discussions. PS gratefully acknowledges the Ministerio de Ciencia e Innovación for the concession of a Juan de la Cierva (JCI-2011-10836) contract and the ICTP, CSIC.

References

1. J. H. Clark and D. J. Macquarrie, *Handbook of Green Chemistry and Technology*, Blackwell Sciences Ltd, Oxford, 2002.
2. B. Hahn-Hagerdal, M. Galbe, M. F. Gorwa-Grauslund, G. Liden and G. Zacchi, *Trends Biotechnol.*, 2006, **24**, 550–556.
3. M. C. Y. Chang, *Curr. Opin. Chem. Biol.*, 2007, **11**, 677–684.
4. A. Demirbas, *Energy Conver. Manag.*, 2003, **44**, 1465–1479.
5. Y. Sun and J. Y. Cheng, *Biores. Technol.*, 2002, **83**, 1–11.
6. S. Czernik and A. V. Bridgwater, *Energy Fuels*, 2004, **18**, 590–598.
7. J. P. Tierney and P. Lidstrom, *Microwave Assisted Organic Synthesis*, ed. J. P. Tierney and P. Lidstrom. Blackwell Publishing: Oxford, 2005.
8. B. L. Hayes, *Microwave Synthesis: Chemistry at the Speed of Light*, CEM Publishing Matthews, NC, 2002.
9. K. Wendler, J. Thar, S. Zahn and B. Kirchner, *J. Phys. Chem. A*, 2010, **114**, 9529–9536.
10. N. Hall, *The New Chemistry*, Cambridge University Press, 2000.
11. D. M. P. Mingos and D. R. Baghurst, *Chem. Soc. Rev.*, 1991, **20**, 1–47.
12. D. E. Clark and W. H. Sutton, *Ann. Rev. Mater. Sci.*, 1996, **26**, 299–331.
13. D. Martin, D. Ighigeanua, E. Mateescua, G. Craciuna and A. Ighigeanu, *Radiat. Phys. Chem.*, 2002, **65**, 63–65.
14. S. A. Galema, *Chem. Soc. Rev.*, 1997, **26**, 233–238.
15. L. Zong, S. Zhou, N. Sgriccia, M. Hawley and L. Kempel, *The Journal of Microwave Power and Electromagnetic Energy: A Publication of the International Microwave Power Institute*, 2003, **38**, 49.
16. T. Razzaq and C. O. Kappe, *ChemSusChem*, 2008, **1**, 123–132.
17. A. Corsaro, U. Chiacchio, V. Pistara and G. Romeo, *Curr. Org. Chem.*, 2004, **8**, 511–538.
18. A. de la Hoz, A. Diaz-Ortiz and A. Moreno, *Chem. Soc. Rev.*, 2005, **34**, 164–178.
19. M. J. Gronnow, R. J. White, J. H. Clark and D. J. Macquarrie, *Org. Process Res. Dev.*, 2005, **9**, 516–518.
20. R. Ruan, P. Chen, R. Hemmingsen, V. Morey and D. Tiffany, *Int. J. Agric. Biol. Eng.*, 2008, **1**, 64–68.
21. Y. F. Huang, W. H. Kuan, S. L. Lo and C. F. Lin, *Biores. Technol.*, 2008, **99**, 8252–8258.
22. A. Loupy, *Microwaves in Organic Synthesis*, 2nd edn., ed. A. Loupy, WILEY-VCH Verlag, Weinheim, 2006.

23. L. Perreux and A. Loupy, *Tetrahedron*, 2001, **57**, 9199–9223.
24. X. Zhang and D. O. Hayward, *Inorg. Chim. Acta*, 2006, **359**, 3421–3433.
25. H. Will, P. Scholz and B. Ondruschka, *Chem. Eng. Technol.*, 2004, **27**, 113–122.
26. X.-j. Bi, P.-j. Hong, X.-g. Xie and S.-s. Dai, *React. Kinet. Catal. Lett.*, 1999, **66**, 381–386.
27. S. J. Vallee and W. C. Conner, *J. Phys. Chem. C*, 2008, **112**, 15483–15489.
28. K. Weissermel and H.-J. Arpe, *Industrial Organic Chemistry*, ed. A. Loupy, WILEY-VCH Verlag, Weinheim, 2007.
29. L. S. Suib and Z. Zang, Low power density microwave discharge plasma excitation energy induced chemical reactions. U.S. Patent No. 5,015,349. 14 May 1991.
30. M. T. Radoiu, Y. Chen and M. C. Depew, *Appl. Catal. B: Environ.*, 2003, **43**, 187–193.
31. K. Onoe, A. Fujie, T. Yamaguchi and Y. Hatano, *Fuel*, 1997, **76**, 281–282.
32. J. A. Menendez, A. Dominguez, Y. Fernandez and J. J. Pis, *Energy Fuels*, 2007, **21**, 373–378.
33. P. Bergese, I. Colombo, D. Gervasoni and L. E. Depero, *Mater. Sci. Eng.: C*, 2003, **23**, 791–795.
34. *U.S. Pat.*, 4,880,578, 1989.
35. *U.S. Pat.*, 4,219,361, 1980.
36. D. E. Clark, D. C. Folz, S. J. Oda and R. Silberglitt, *Microwaves: Theory and Application in Materials Processing III*, American Ceramic Society, 1995.
37. B. Krieger, Proceedings of the American Chemical Society, Division of Polymeric Materials: Science and Engineering, 1992.
38. F. M. Thuillier and H. Jullien, *Makromolekulare Chemie. Macromolecular Symposia*, 1989, **25**, 63–73.
39. S. Komarneni, Q. Li, K. M. Stefansson and R. Roy, *J. Mater. Res.*, 1993, **8**, 3176–3183.
40. C. O. Kappe, *Angew. Chem. Int. Ed.*, 2004, **43**, 6250–6284.
41. R. Luque, S. K. Badamali, J. H. Clark, M. Fleming and D. J. Macquarrie, *Applied Catalysis A: General*, 2008, **341**, 154–159.
42. R. S. Varma, A. K. Chatterjee and M. Varma, *Tetrahedron Lett.*, 1993, **34**, 3207–3210.
43. R. S. Varma and R. Dahiya, *Tetrahedron Lett.*, 1997, **38**, 2043–2044.
44. A. Domínguez, J. A. Menéndez, Y. Fernández, J. J. Pis, J. M. V. Nabais, P. J. M. Carrott and M. M. L. R. Carrott, *J. Anal. Appl. Pyrolysis*, 2007, **79**, 128–135.
45. T. Bridgwater, *J. Sci. Food Agric.*, 2006, **86**, 1755–1768.
46. R. Fahmi, A. V. Bridgwater, I. Donnison, N. Yates and J. M. Jones, *Fuel*, 2008, **87**, 1230–1240.
47. G. G. Allan, B. B. Krieger and D. W. Work, *J. Appl. Polym. Sci.*, 1980, **25**, 1839–1859.

48. R. W.-C. Chan and B. B. Krieger, *J. Appl. Polym. Sci.*, 1981, **26**, 1533–1553.

49. M. Miura, H. Kaga, A. Sakurai, T. Kakuchi and K. Takahashi, *J. Anal. Appl. Pyrolysis*, 2004, **71**, 187–189.

50. M. Miura, H. Kaga, T. Yoshida and K. Ando, *J. Wood Sci.*, 2001, **47**, 502–506.

51. A. M. Sarotti, R. A. Spanevello and A. G. Suarez, *Green Chem.*, 2007, **9**, 1137–1140.

52. A. Domínguez, J. Menéndez, M. Inguanzo and J. Pis, *Fuel Process. Technol.*, 2005, **86**, 1007–1020.

53. A. Domínguez, J. A. Menendez, M. Inguanzo and J. J. Pis, *Biores. Technol.*, 2006, **97**, 1185–1193.

54. F. Yu, P. Chen, A. I. Tunheim, X. Lin, Y. Liu and R. Ruan, 28th Symposium on Biotechnology for Fuels and Chemicals, 2006.

55. F. Yu, S. P. Deng, P. Chen, Y. Liu, Y. Wang, A. Olsen, D. Kittelson and R. Ruan, *Applied Biochemistry and Biotechnology*, 2007, **136–140**, 957–970.

56. B. C. Saha, in *Lignocellulose Biodegradation*, American Chemical Society, Washington, D.C., 2004, pp. 2–34.

57. V. L. Budarin, J. H. Clark, B. A. Lanigan, P. Shuttleworth, S. W. Breeden, A. J. Wilson, D. J. Macquarrie, K. Milkowski, J. Jones, T. Bridgeman and A. Ross, *Biores. Technol.*, 2009, **100**, 6064–6068.

58. R. Luque, *Energy Environ. Sci.*, 2010, **3**, 254–257.

59. A. B. Ross, J. M. Jones, M. L. Kubacki and T. Bridgeman, *Biores. Technol.*, 2008, **99**, 6494–6504.

60. S. Wang, X. M. Jiang, N. Wang, L. J. Yu, Z. Li and P. M. He, *Energy Fuels*, 2007, **21**, 3723–3729.

61. D. Muraoka, *Bull. Fish. Res. Agency Jpn.*, 2004, 59–64.

62. V. L. Budarin, Y. Zhao, M. J. Gronnow, P. S. Shuttleworth, S. W. Breeden, D. J. MacQuarrie and J. H. Clark, *Green Chem.*, 2011, **13**, 2330–2333.

63. V. Ferreira-Leitão, L. Gottschalk, M. Ferrara, A. Nepomuceno, H. Molinari and E. S. Bon, *Waste Biomass Valor*, 2010, **1**, 65–76.

64. W. Widmer, W. Zhou and K. Grohmann, *Biores. Technol.*, 2010, **101**, 5242–5249.

65. J. Lopez, L. Qiang and I. Thompson, *Biotechnology*, 2010, **30**, 63–69.

66. A. M. Balu, V. Budarin, P. S. Shuttleworth, L. A. Pfaltzgraff, K. Waldron, R. Luque and J. H. Clark, *ChemSusChem*, 2012, **5**, 1694–1697.

67. C. S. K. Lin, L. A. Pfaltzgraff, L. Herrero-Davila, E. B. Mubofu, S. Abderrahim, J. H. Clark, A. A. Koutinas, N. Kopsahelis, K. Stamatelatou, F. Dickson, S. Thankappan, Z. Mohamed, R. Brocklesby and R. Luque, *Energy Environ. Sci.*, 2013, **6**, 426–464.

68. L. A. Pfaltzgraff, M. De Bruyn, E. C. Cooper, V. Budarin and J. H. Clark, *Green Chem.*, 2013, **15**, 307–314.

69. J. H. Clark, E. M. Fitzpatrick, D. J. Macquarrie, L. A. Pfaltzgraff and J. Sherwood, *Catal. Today*, 2012, **190**, 144–149.

70. T. Durka, G. D. Stefanidis, T. Van Gerven and A. Stankiewicz, *Meas. Sci. Technol.*, 2010, **21**.

71. *The Exploitation of Pyrolysis Oil in the Refinery Main Report* 40661, The Carbon Trust, 2008.

72. R. J. Evans and T. A. Milne, *Energy Fuels*, 1987, **1**, 123–137.

73. D. J. Nowakowski, J. M. Jones, R. M. D. Brydson and A. B. Ross, *Fuel*, 2007, **86**, 2389–2402.

74. B. M. Jenkins, L. L. Baxter, T. R. Miles Jr and T. R. Miles, *Fuel Process. Technol.*, 1998, **54**, 17–46.

75. D. J. Nowakowski and J. M. Jones, *J. Anal. Appl. Pyrolysis*, 2008, **83**, 12–25.

76. R. Fahmi, A. V. Bridgwater, S. C. Thain, I. S. Donnison, P. M. Morris and N. Yates, *J. Anal. Appl. Pyrolysis*, 2007, **80**, 16–23.

77. D. J. Nowakowski, C. R. Woodbridge and J. M. Jones, *J. Anal. Appl. Pyrolysis*, 2008, **83**, 197–204.

78. C. Di Blasi, C. Branca and A. Galgano, *Polym. Degrad. Stabil.*, 2008, **93**, 335–346.

79. Y. Tang, W. Yu, L. Mo, H. Lou and X. Zheng, *Energy Fuels*, 2008, **22**, 3484–3488.

80. X. Junming, J. Jianchun, S. Yunjuan and L. Yanju, *Biomass Bioenergy*, 2008, **32**, 1056–1061.

81. Q. Zhang, J. Chang, T. Wang and Y. Xu, *Energy Conver. Manag.*, 2007, **48**, 87–92.

82. D. Mu, T. P. Seager, P. S. C. Rao, J. Park and F. Zhao, *Int. Environ. Assess. Manag.*, 2011, **7**, 348–359.

Food Wastes Conversion to Products for Use in Chemical and Environmental Technology, Material Science and Agriculture

ENZO MONTONERI,* ALESSANDRA BIANCO PREVOT,[a]
PAOLA AVETTA,[a] ANTONIO ARQUES,[b]
LUCIANO CARLOS,[c] GIULIANA MAGNACCA,[a]
ENZO LAURENTI[a] AND SILVIA TABASSO[a]

[a] Università di Torino, Dipartimento di Chimica, Via P. Giuria 7, 10125 Torino, Italy; [b] Grupo de Procesos de Oxidación Avanzada, Dpto de Ingeniería Textil y Papelera, Universidad Politécnica de Valencia. Plaza Ferrándiz y Carbonell s/n, Alcoy, Spain; [c] Instituto de Investigaciones Fisicoquímicas Teóricas y Aplicadas (INIFTA), Diag 113 y 64, La Plata, Argentina
*Email: enzo.montoneri@unito.it

4.1 Introduction

The economic utilisation of food coproducts cannot ignore the problem of managing food wastes. While being the current matter of concern for the environmental and economic impact connected to their disposal, these wastes contain chemical energy that can in principle be recovered and recycled for

RSC Green Chemistry No. 24
The Economic Utilisation of Food Co-Products
Edited by Abbas Kazmi and Peter Shuttleworth
© The Royal Society of Chemistry 2013
Published by the Royal Society of Chemistry, www.rsc.org

further use. Indeed, the production of thermal and electric energy using the chemical energy contained in the organic matter of biomasses is a current strategic goal to reduce dependence on fossil fuels.[1] Current technologies used for this purpose have, however, a major critical point in the low conversion of the biomass to potential chemical energy, as discussed below.

There are basically two ways to convert biomass chemical energy to heat and/or electricity, *i.e.* combustion and fermentation. Biomass combustion is expensive due to the high content of water. In addition, incineration plants are restricted by the public concern over the release of fine dust particles, and by the very costly flue-gas treatment processes and equipment. Fermentation allows more concentrated and cleaner fuels to be obtained, such as methane and ethanol. However, this process suffers due to low biomass conversion. Indeed, no more than 50% of the starting organic matter is converted to the desired fuel product. In addition, fermentation processes lead to organic N mineralisation. This poses the problem of disposing of the residual organic matter for its environmental impact.[2] The final result of all these factors is that the cost of the obtained fuels is greater than their value. The main reason for the low yield of fermentation processes is that the available micro-organisms metabolise mostly the carbohydrate fraction of biomasses, and have limited product tolerance, and are not able and/or are inhibited by the lignin fraction.[3] Improvements to the fermentation process are sought through selection of micro-organisms' consortia or creating by biotechnology novel enzymes with higher substrate/product tolerance, and/or through development of new biomass pretreatment procedures to separate cellulose from lignin. The aim is to increase product yield and concentration in the fermentation liquor and substrate utilisation efficiency.[4] Aside from this, it is unanimously recognised that biofuel production plants can become cost effective only by coupling the fuel production process to the treatment of the residual organic lignin fractions in order to obtain high added value products to sell to the consumer's market. This raises the level of research development from the mere production of biofuels to the more ambitious development of a biorefinery fed with biomasses to produce fuel, chemicals and consumer products.[5,6]

A second major issue in the production of renewable energy is the use of biomass from dedicated crop *versus* biorefuse. While the former raises ethical concerns, the latter allows two important current issues to be addressed at the same time, *i.e.* the use of renewable energy and the management of wastes. A proper exploitation of biomass energy will automatically solve biowaste-management problems.

The sustainable conversion from a fossil-fuel-based economy to a renewable-energy economy is not dependent on the development of new technology as much as on the entropy level of the energy source. The exploitation of fossil-fuel sources has allowed the development of enough technology that is now available for use in the exploitation of renewable-energy sources. These latter, however, suffer from being spread over large areas, which primarily impose high collection costs. Urban wastes are an exception. The environmental impact of urban wastes has dramatically increased as a result of increasing population urbanisation and consumption habits. Although nowadays this fact contributes to society a significant economic burden for management and/or disposal, concentration of wastes in urban areas has in turn allowed the collection of natural bio-organic matter in well-confined spaces. The present chapter reports how, through urbanisation and municipal collection practices, urban biowastes have become a low-entropy potential sustainable source of both energy and materials for the chemical industry and society.

4.2 Food Wastes as a Source of Products for the Chemical Industry

The Department of Chemistry of the University of Torino, within the Biochemenergy project, has shown that the recalcitrant lignin-like fraction of urban biowastes (UBW) is a cost-effective source of chemical auxiliaries[7] that can find application in the formulation of detergents, textile dyeing baths, emulsifiers, auxiliaries for soil/water remediation, flocculants, dispersants and binding agents for ceramics manufacture, and nanostructured materials for chemical and biochemical catalysis. A patent for both process and products has been recently filed.[8] A range of biowastes available in metropolitan areas have been investigated. These were sampled from a waste-treatment plant located in the Piemonte region of Italy. This installation treats the biowastes produced by a population of 2.9 millions spread over an area 13 732 km^2, with 30% people living in one large metropolitan area and the rest distributed over other 564 small municipalities.[9] The investigated wastes were the organic humid fraction of urban food wastes (UFW), home gardening and public park trimming residues (UGR), sewage sludge (SS) from municipal wastewater treatment, and their fermentation products after anaerobic or aerobic digestion. These materials were hydrolysed in alkaline water to yield a liquid/solid mix that was allowed to settle in order to separate the supernatant liquid phase containing the soluble hydrolysed from the insoluble residue (IOR). The recovered liquid phase was circulated through a polysulfone ultrafiltration membrane with 5 kD molecular weight cut-off to yield a retentate with 5–10% dry matter content. The membrane retentate was dried at 60 °C to yield the final water soluble biobased product (SBO).

Investigation of the chemical nature of SBO has shown that these substances bear chemical similarities with natural soil and water humic substances (HS). Depending on the sourcing materials, the SBO have been found to be mixtures of molecules differing in molecular weight (MW) from 67 to 463 kg mol^{-1}, and in the C-type, content and functional groups. They are described as likely mixtures of substances formed by long aliphatic carbon chains substituted by aromatic rings and functional groups such as COOH, CON, C=O, PhOH, O_alkyl, O_aryl, OCO, OMe, and NRR', (R and R' = alkyl or H). These organic moieties represent the memory of fats, proteins, polysaccharides and lignin contained in the sourcing refuse matter that is not completely mineralised by biodegradation. For this reason, SBO may be considered to be the pristine material of natural soil organic matter formed under longer aging conditions. Thus, the structural similarities between SBO and HS are not surprising.

Due to the complex chemical composition, the behaviour of SBO in water solution cannot be understood as well as for synthetic single molecules. At low concentration, the SBO have been found to behave as small-molecule surfactants. They give water surface tension (γ) *versus* added SBO concentration (C_S) plots decreasing steeply upon increasing C_S until a slope change occurs to a more or less flat plateau at $1–2\,g\,L^{-1}$ C_S. Their surface activity properties correlate with their power to enhance the solubility of hydrophobic compounds in water. At $50–100\,g\,L^{-1}$ C_S they behave more like water soluble polyelectrolytes; they are no longer soluble in water and yield viscous gel-like phases separating from the bulk water phase.[7]

The surfactant properties of SBO play a major role to determine their performance as detergents or auxiliaries for textile dyeing. By virtue of their capacity to lower water surface and interfacial tension they favor the water contact with dirty surfaces. At the same time, they have the capacity to form micelles or to acquire pseudomicellar conformation in the bulk water phase, and are capable of enhancing the water solubility of hydrophobic materials. Therefore, they can be used to help the transfer of dirt from objects to be washed into the water phase. This process has been demonstrated to be effective in washing solutions containing SBO as well as for washing solutions containing synthetic surfactants.[10] A similar mechanism occurs in dyeing textiles. In this case, the SBO have been shown to be capable of binding water-soluble or -insoluble dyes, and modulate their transfer kinetics to the surface of the fabric to be dyed, thus yielding fabrics dyed as intensely and homogeneously as synthetic surfactants.[11]

The capacity of the SBO molecules to acquire different conformation and/or yield large macromolecular aggregates in solution has offered scope to investigate these substances as templates for the synthesis of nanostructured materials. Also, their structural similarity with HS, and the properties of HS as photosensitisers and their role in soil fertility, has stimulated investigation of SBO for specific applications in environmental remediation and agriculture.

4.3   Food Wastes as Source of Products for Material Science

The behaviour of SBO in water solution makes them a suitable medium to obtain materials for technologic applications, in particular to obtain high surface area, porous materials.

Among several synthetic strategies, we can recall:

1) sol-gel synthesis using templating agents;
2) monolith production from preformed oxidic particles.

4.3.1 Sol-Gel Synthesis (Using Templating Agents)

Porous materials with large specific surface areas, extended porous networks and controlled pore-size distributions can be obtained using a sol-gel synthesis assisted by surfactant molecules. All these features are particularly important dealing with materials for (photo)catalysis, adsorbents and others whose high surface area is fundamental for their applications.

Sol-gel synthesis initiates from precursors that can hydrolyse and transform in the presence of an initiator and/or a catalyst, allowing the formation of a gel-phase. The final product begins to form in the shape of spheres or chains, depending on the experimental conditions applied. Once the product is formed, the solvent is eliminated in the subsequent drying phase, and the final calcination phase brings about the formation of a material with chemical/thermal stability and good mechanical properties.

In sol-gel synthesis, SBO molecules can act in two directions: (i) as templating agents, and (ii) as nucleation directing agents.

4.3.1.1 SBO as Templating Agent

Since SBO molecules are formed by polar groups and apolar fragments, they can act as biosurfactants and, in this respect, they can form micelles, vescicles and other supramolecular aggregates, depending on their concentration in aqueous solution. The preferred aggregation shape at a given concentration can drive the polymerisation of the oxide precursor in a sol-gel process forming porous oxidic materials with the chosen porosity.

A typical surface tension *vs.* SBO aqueous solution concentration curve is reported in Figure 4.1.[12]

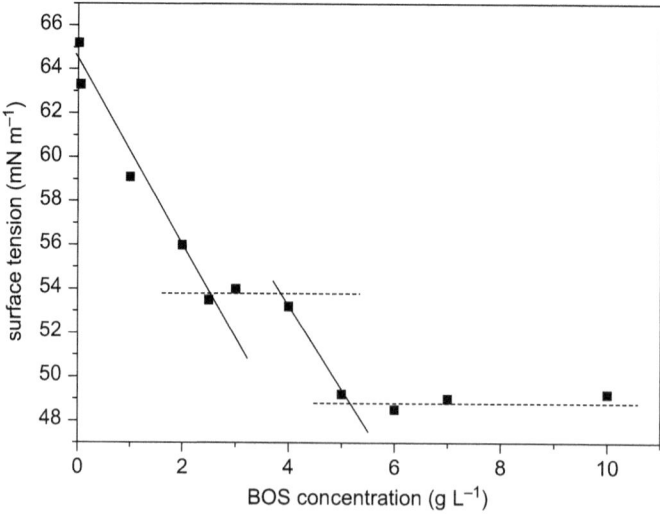

Figure 4.1   Plot of surface tension (γ) *vs.* SBO (type BS110) concentration in water at 25 °C and pH 5.

The plot evidences two steeply descending tracts (solid lines) and two plateaus (broken lines). Similar behaviour is shown by other surfactant-like molecules.[13] The two observed plateaus arise from the first aggregation of single surfactant molecules into micellar aggregates, and from micellar aggregates that convert into vesicles at higher SBO concentrations.

This result was expected to allow the fabrication of different porous materials simply by varying SBO concentration in the reaction medium. In the following, the synthesis of mesoporous silica is described.

Usually, the synthesis of silica particles in the presence of anionic surfactants occurs at pH < 2, but at this pH SBO is not soluble. To solve this problem, TEOS was used together with 3-aminopropyltriethoxysilane (APS) that facilitates the process to operate at pH 5 with the surfactant in solution. Under these conditions the sol-gel reaction takes place and the silica network forms following the shape of the biosurfactant molecules or aggregates. Samples are indicated as SS0, SS3 and SS10 considering the concentration of SBO solutions employed (0 g/L, 3 g/L and 10 g/L, the latter two concentrations in the range relative to formation of micellar and vescicular aggregates, respectively).

The porous materials obtained were characterised *via* HRTEM and N_2 gas-volumetric adsorption at 77 K. Micrographs and the pore-size distribution curves are reported in Figures 4.2 and 4.3, respectively.

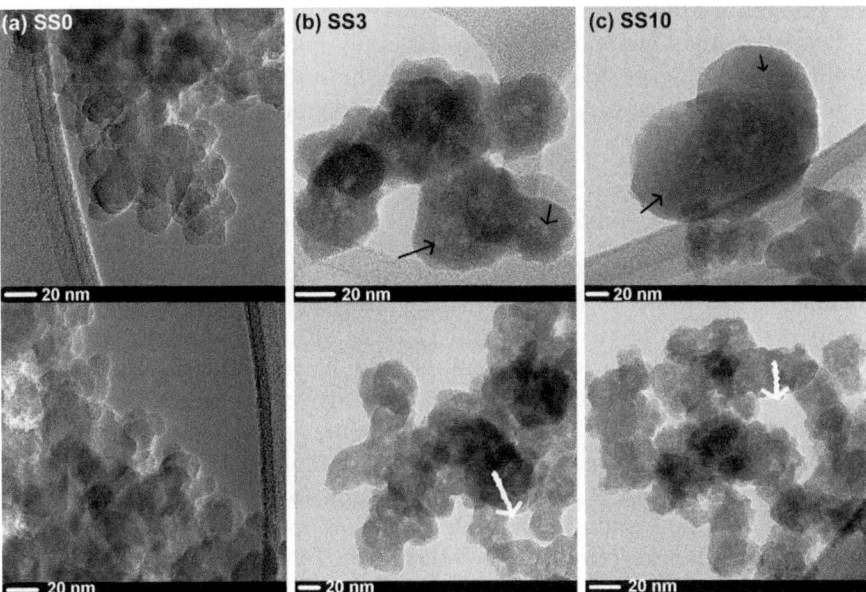

Figure 4.2 TEM images of calcined silica powders: SS0 (a), SS3 (b) and (c) SS10. The arrows in the pictured indicate intraparticle cavities (black arrows) and interparticles cavities (white arrows).
Reproduced with permission from Wiley-VCH Verlag GmbH & Co. KGaA, Weinheim.

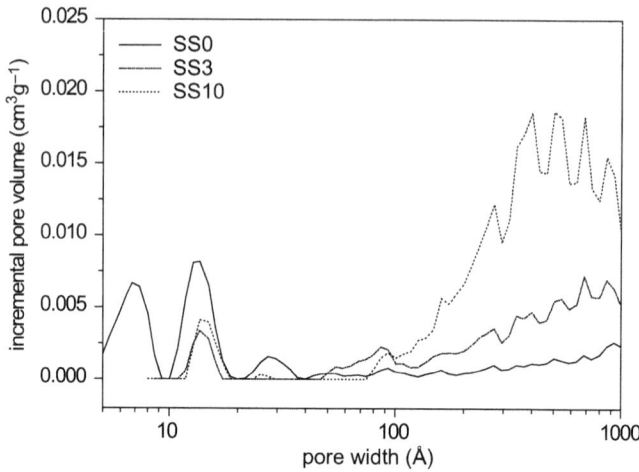

Figure 4.3   Pore-size distributions of the SS0, SS3 and SS10 samples. The pore-size distributions were obtained by applying the DFT method (slit pores, low regularisation).

Both analyses evidence the presence of micropores (~ 7 and ~ 15 Å of width) in SS0 samples, whereas the pore size becomes larger and larger on increasing the amount of SBO employed.

It is clear that mesoporous silica materials may be obtained *via* sol-gel synthesis; by using waste-derived polymeric biosurfactant as templates, their pore-size distributions can be tuned by varying the biosurfactant concentration in the reaction medium.

4.3.1.2   SBO as Nucleation Directing Agent

In 1990 Patel and Padhi studied the possibility of using jute fibres to produce alumina fibres.[14] Since then, lots of papers have been published dealing with the use of natural substances in developing green sustainable materials (nanostructured inorganic particles, monoliths and films).[15–31] The utilisation of natural substances as structure-directing agents is very attractive because they possess specific reactivity, allow the fabrication of complex inorganic structures following easy synthetic pathways, and are renewable and often inexpensive.

In this context, waste-derived biosurfactants (SBO) were also tested as structure-directing agents in the preparation of mesoporous TiO_2 powders starting from titanium tetraisopropoxide (TIIP).[32] Different aqueous solutions of SBO were employed, namely 0, 10 and 50 g/L. All the samples (indicated as T00, T10 and T50) were characterised in order to evidence morphology and structure modifications induced by SBO presence.

Table 4.1 reports the specific surface area of the three samples, the total amount of porosity and pore width (also reported in Figure 4.4) and the crystal size as evaluated by the Scherrer equation from 4RD data (calculated from

Table 4.1 Morphological behaviours of TiO$_2$ powders.

Sample	BET area $(m^2 g^{-1})$	Total amount of mesopores $(cm^3 g^{-1})$	Pore width $(\mathring{A})$	Crystal size^a $(\mathring{A})$
T00	88	0.42	200	165
T10	80	0.30	100	95
T50	69	0.19	70	80

^{a}Calculated from signal at $2\theta = 29.5$ relative to (101) plane of TiO$_2$ anatase, shape factor $= 0.9$.

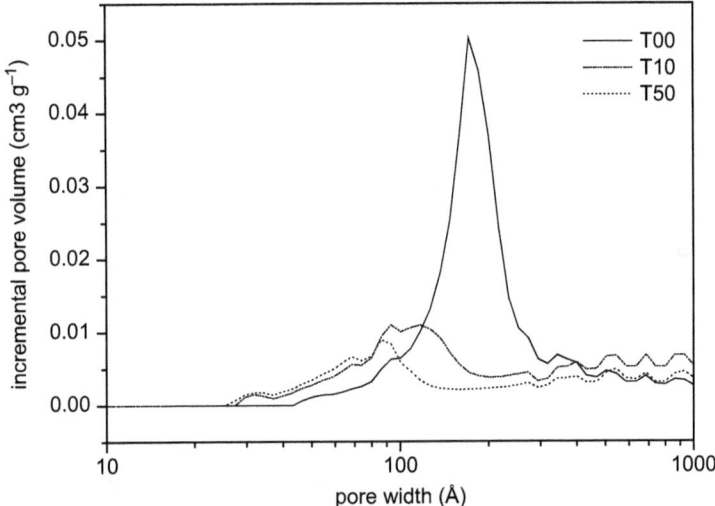

Figure 4.4 Pore-size distributions of the T00, T10 and T50 samples. The pore-size distributions were obtained by applying the DFT method (slit pores, low regularisation).

signal at $2\theta = 29.5$ relative to the (101) plane of TiO$_2$ anatase, shape factor $= 0.9$).

Figure 4.4 shows the pore-size distribution obtained for the three TiO$_2$ samples.

Nitrogen adsorption analyses show that the use of increasing amounts of SBO in the synthesis procedure leads to a material that is more dense (less porous) and is characterised by larger particles. Otherwise, the size of crystalline domains decreases, as indicated by 4RD data. In this case, SBO molecules do not act as a template, but rather seem to affect the nucleation and particle growth kinetics.

A possible explanation for this phenomenon is the presence of polar functional groups, such as phenol and carboxylic groups, in the SBO molecules. In fact, polar groups may interact *via* hydrogen bonding with inorganic precursors, affecting the particles growth and consequently their size. This behaviour was discussed recently by He *et al.* in a paper reporting yeast cell assisted synthesis of a photoactive TiO$_2$ powder.[26] During fermentation, the yeast cells produced macromolecules with surfactant properties including proteins and polysaccharides. The anionic hydrophilic groups on these macromolecules (mainly $-COO^-$ and $-OPO_3^{2-}$) provided oriented nucleation sites for the positively charged Ti (IV) ions during the sol synthesis. In the presence of SBO molecules, COO^- and OH can interact with the electrophilic metal centres forming Ti–O–SBO adducts, where O–SBO represents a biosurfactant molecular fragment. Then, these Ti–O–SBO centres can act as nucleation sites, around which titanium nanoparticles develop *via in situ* hydrolysis and condensation of TIIP. Increasing the number of nucleation sites by increasing SBO concentration in the synthesis pot, the TiO_2 particles growth occurs faster and the crystallinity degree is limited. This is consistent with the decrease observed in the specific surface areas and with the smaller crystal sizes evidenced by the T10 and T50 samples X-ray diffractograms. The same effect given by polar organic groups can be given by the polar inorganic residues derived from the complex starting waste composition (revealed by elemental analyses), and/or used during the separation/purification of the biomolecule (acidification/basification processes): both residues allow the SBO molecules to act as nucleation-directing agents.

4.3.2 Monoliths Production from Preformed Oxidic Particles

Monoliths of preformed SiO_2 particles were prepared, and physicochemically characterised, using SBO as a binder,[33] taking inspiration from the glass-frit bonding technique.[34] Solder glasses were specially designed for joining glass to other glasses, ceramics, or metals at low temperatures (not higher than 450 °C) and excellent results were obtained over time in various application fields (electronics, solar cells, packaging and others[35–37]). The mixture used for the glass-frit procedure is made up of silica, organic binders and inorganic fillers: together they are able to form a molten layer at low temperatures (450 °C) that constitutes the connection between different materials. The mixture described above, fits well with the SBO composition, where inorganic moieties derived from waste sources and/or from chemical treatments that are carried out during the purification/separation stages. Figure 4.5 reports a SEM image of a monolith. It shows very smooth, extended surfaces, whereas the planes created by breaking the monolith are much more irregular, suggesting the presence of aggregates of large dimensions. This indicates that the crosslinking between silica particles is limited and does not lead to the formation of a bulky silica block.

To confirm the presence of void spaces inside the monolith, N_2 adsorption measurements at 77 K were performed on two monoliths, obtained from nonporous and porous silica. The main features are reported in Table 4.2.

The specific surface area of the monoliths are lower than the relative powders, indicating a sintering of the starting particles. Monoliths formed using a nonporous silica develop an extended porosity, whereas the porosity already present in the porous silica material are slightly modified by monolith formation that produces smaller pores leading to a smaller pore volume.

These data indicate that the procedure just described can be applied to the production of handy and cheap devices where the presence of large available

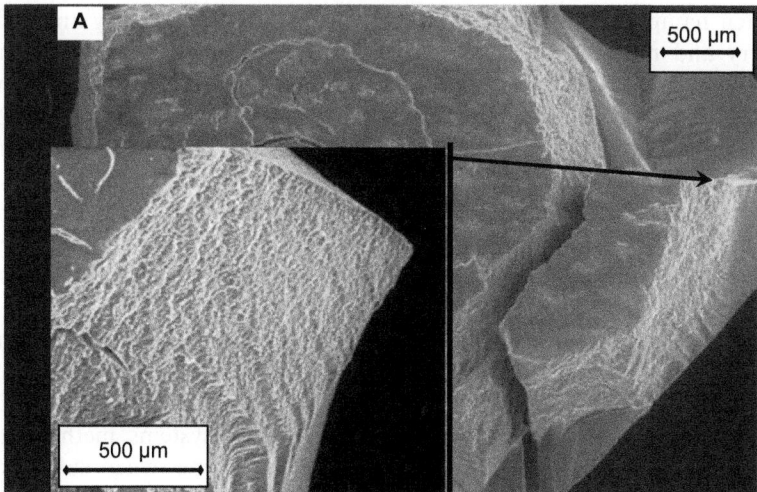

Figure 4.5 SEM images of silica-monolith (in the insert: a detail of the plane observed after monolith break).

Table 4.2 Morphological behaviours of powder and glass-frit treated silicas.

Sample	BET area ($m^2\ g^{-1}$)	BJH Total amount of pores ($cm^3\ g^{-1}$)	BJH Pore width ($\mathring{A}$)
Nonporous silica	297	–	–
Nonporous silica monolith	198	1.17	200
Porous silica	164	1.61	400
Porous silica monolith	124	0.98	330

surface and open porosity is required. For instance, an interesting application of the prepared monolith is as support for the production of biocatalysts that is reported in the next section.

4.3.3 SBO Usage in Biocatalysis

As affirmed by Sheldon: "In the drive towards green, sustainable metho-dologies for chemicals manufacture biocatalysis has much to offer".[38] Indeed, catalysts based on living micro-organisms or enzymes allow high activities and selectivities to be obtained, whilst using mild reaction conditions (pressure, pH and temperature) in environmentally tolerable solvents (or directly in water). Moreover, modern biotechnology techniques makes real two possibilities: to produce high quantities of enzymes at a commercially acceptable price, and to manipulate their original structure in order to obtain new properties (this is representative of new synthetic routes, which are surely more attractive than traditional organic syntheses).

Nevertheless, to use the enzymes on an industrial scale, some important issues need to be solved because, again citing Sheldon: "industrial applications are often hampered by a lack of long-term operational stability and difficult recovery and reuse of the enzyme." These drawbacks can be overcome by enzyme immobilisation on a suitable material able to ensuring the maintenance of catalytic activity over time. From this point of view, silica monoliths obtained by SBO-template method have the potential required for this kind of material: stability, good mechanical properties and surface features adjustable by choosing different silica or appropriate pretreatments. Furthermore, these materials can be easily functionalised to favour enzyme–monolith interactions or, eventually, formation of covalent bonds between enzyme molecules and support.

There are several reasons for using an enzyme in immobilised form: (i) more convenient enzyme handling; (ii) easier separation from the products, thereby minimising protein contamination of the final products; (iii) more efficient recovery and reuse of the enzyme; (iv) continuous systems methods to be designed and used; (v) enhanced stability of the enzyme, when submitted to denaturation by heat, organic solvents, or autolysis.

In recent years, many different methods have been reported for the immobilisation of enzymes. These methods can be summarised as: adsorption, entrapment, crosslinking, and covalent immobilisation (Figure 4.6).

The immobilisation by adsorption is the most used and reliable method, but generally the interaction that takes place between the enzyme molecules and the surface groups are too weak to be stable for a long time under reaction conditions, so other techniques have been studied in order to obtain more stable and resistant systems.[39] The entrapment method is an improvement of the previous one and it generally consists in the inclusion of the enzyme in a polymer network or in a sol-gel system. Moreover, in some studies, enzymes were also trapped within a porous membrane whose pores are large enough to allow the passage of substrates and reaction products, but not the removal of protein molecules.[40]

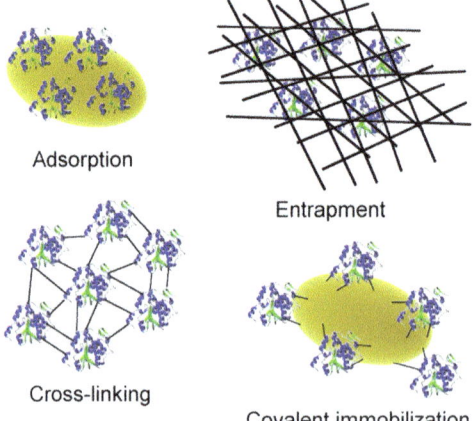

Figure 4.6   Different methods for enzyme immobilisation.

The crosslinking technique consists in the formation of large enzyme aggregates by reaction with bifunctional reagents like glutaraldehyde. The final products are macroparticles easily separable from the reaction batch, but usually a loss of enzyme activity is observed, particularly at high enzyme concentrations. Moreover, with this method, reactivity is poorly reproducible and the biocatalyst shows low mechanical stability, and gelatinous aggregates are difficult to handle. Nevertheless, some of the difficulties can be solved considering that mechanical stability and ease of handling could be improved by crosslinking the enzyme in a gel matrix or on a carrier.

Finally, covalent immobilisation involves the formation of stable chemical bonds between the support and the protein molecules and it usually leads to the production of highly stable systems, even if often the enzyme loses part of its activity. The close interaction between the support and the protein can, indeed, lead to distortions of the three-dimensional structure of the protein or the partial occlusion of its active site. In each case, the balance between these two factors – stabilisation and loss of activity – determines if the best immobilisation strategy has been selected.

In the case of SBO-assisted production of siliceous monoliths, both the adsorption and the covalent methods can be applied. We report in the following some results obtained by covalently immobilising the Soybean Peroxidase (SBP), a stable and versatile enzyme able to catalyse the oxidation of organic and inorganic substrates by hydrogen peroxide.[41] A sketch of the SBP structure (from RSCB Protein Data Bank Id. 1FHF[42]) and a scheme of its catalytic cycle,[43] are reported in Figures 4.7a and 4.7b, respectively.

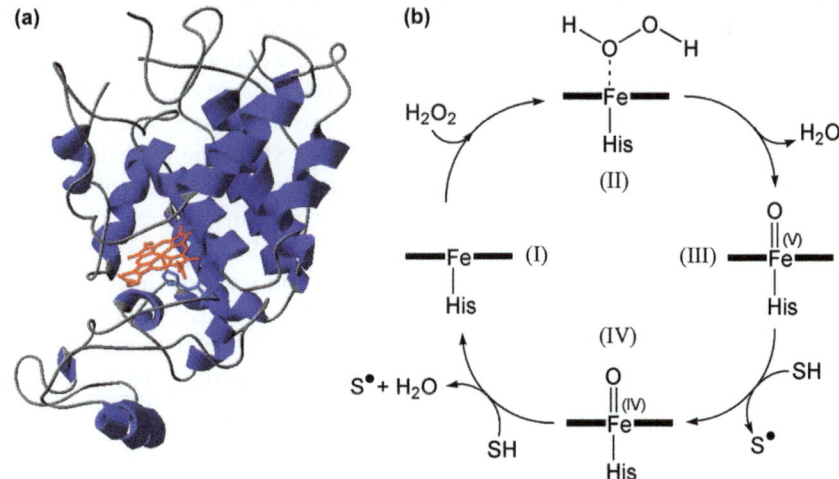

Figure 4.7   Soybean peroxidase: (a) image based on the X-ray,[42] in red the catalytic site (Fe-heme group); (b) catalytic cycle (SH, substrate). The native form of the enzyme (I) reacts with the peroxide molecule (II) giving the first intermediate (III); the oxidisation of the first molecule of substrate generates the second intermediate (IV), which is able to oxidise.

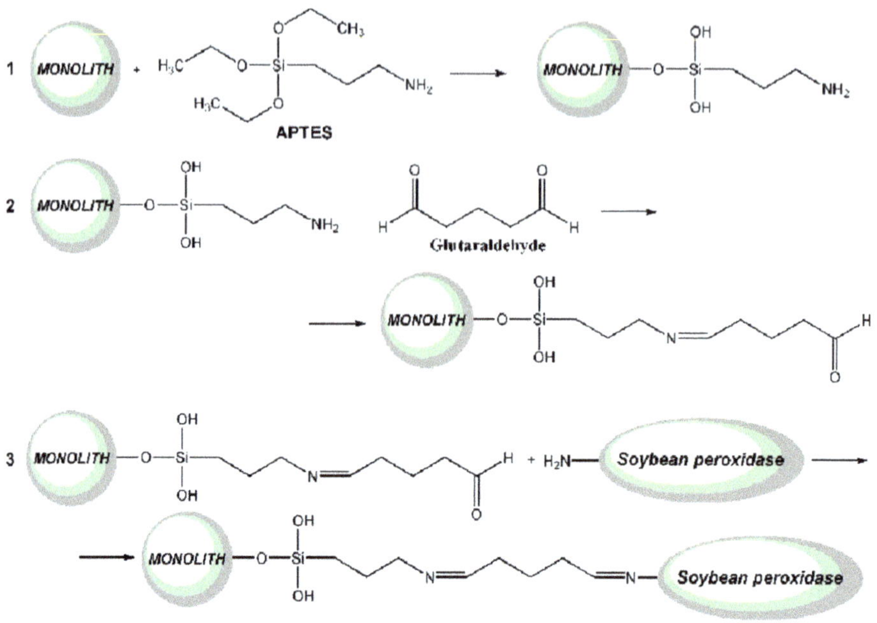

Figure 4.8   Reaction scheme for the immobilisation of SBP on silica.

SBP can be immobilised onto SBO-prepared monoliths by using a method previously applied in the functionalisation of silica beads.[35,44] The support is initially activated by reaction with 3-aminopropyltriethoxysilane in water. Then, as reported in Figure 4.8, SBP was bound to the monoliths by reaction with the homobifunctional linker glutaraldehyde. The covalent bonds resulting from the nucleophilic addition of the aldehyde groups to the amino groups present on the monolith and on the enzyme surface, respectively, ensure the formation of a stable adduct, which can be dried and stored in a refrigerator at 4 °C.

Following the described method, about 11 mg of SBP were loaded for each gram of monolith, whereas 16 mg of SBP were loaded for each gram of powder, but the real loadings, considering the different specific surface area of the two materials, are very similar and correspond to 0.09 mg SBP/m^2 of support.

Both the immobilised SBP on powder (P-SBP) and on monoliths (M-SBP) were tested for their catalytic activity by using the DMAB-MBTH reaction (Figure 4.9a), a colorimetric test usually employed in the determination of peroxidase activity.[45] The results of the activity measurements were reported in Figure 4.9b, together with the data obtained in the absence of SBP and in presence of the free protein in solution.

The experimental data demonstrate that, as expected, the immobilised SBP shows a lower activity than the free enzyme. Otherwise, it is noteworthy that, after 2 h, the SPB immobilised onto the monoliths, achieves the same reagent conversion. On the contrary, the SBP immobilised onto the silica powder seems to have a lower efficiency in the DMAB-MBTH conversion, in fact P-SBP does not exceed 75% conversion of the reagent even after 4 h of

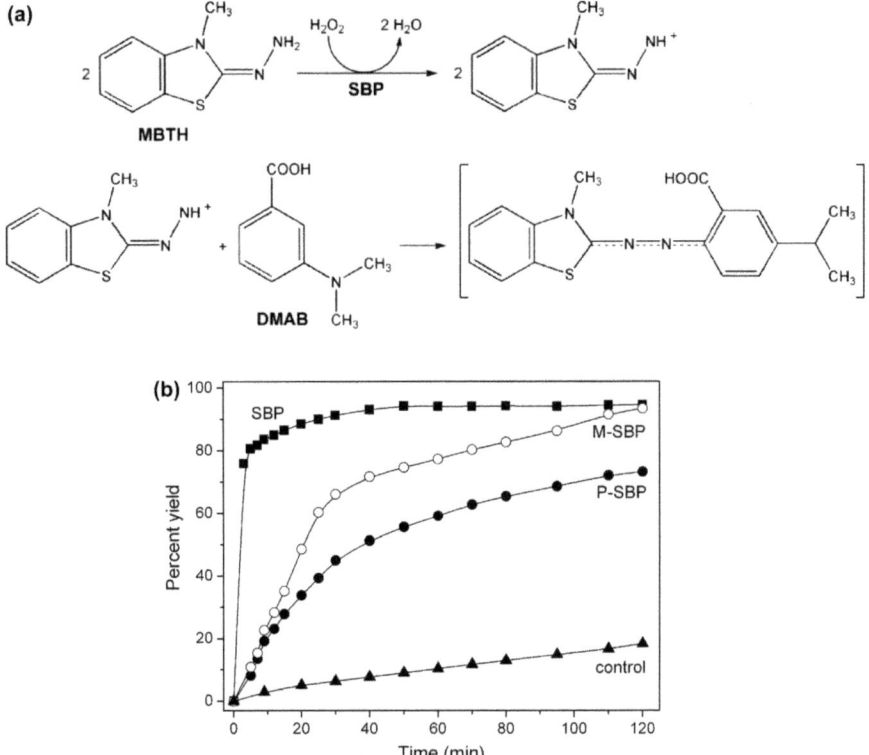

Figure 4.9 (a) Scheme of the DMAB-MBTH reaction; (b) DMAB-MBTH conversion in the absence (control) and presence of SBP, either free or immobilised on silica powder (P-SBP) and monolith (M-SBP).[33] Reproduced with permission from Elsevier.

reaction (data not shown). This phenomenon seems to point to the importance of the mesoporous network, where reagents and biocatalyst can meet and react very easily.

To test performance stability upon repeated recovery and recycling, the immobilised catalysts were reused 20 times. The results showed that, after an initial decline, the activity tends to remain constant at a value corresponding to about 75–80% of the initial one.[33]

The immobilisation onto the monoliths also improves the retention of SBP activity over time. As shown in Figure 4.10, the SBP-monolith samples retain approximately unchanged their catalytic properties for about 40 days (refrigerator at 4 °C). In the same period, the SBP stored in solution completely loses its activity. Moreover, the immobilised protein maintains about one third of its initial activity also after four months.

The potential of these systems in the degradation of pollutants were tested against real wastewater obtained from the exhausted bath of a textile company. The initial solution that contained several substances and a mixture of

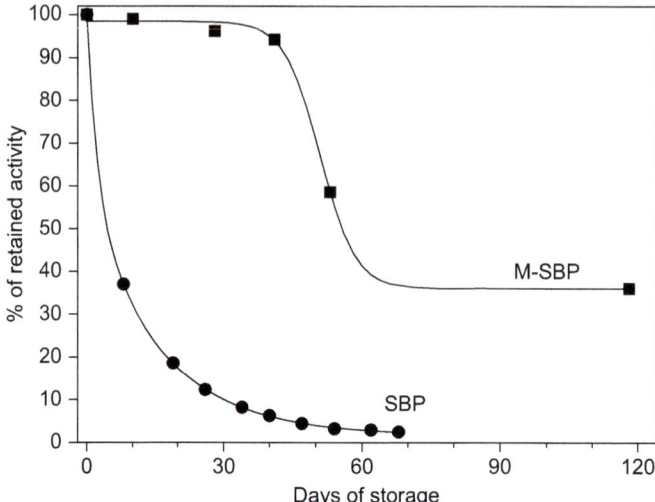

Figure 4.10   Effects of storage at 4 °C on the free and immobilised soybean peroxidase.

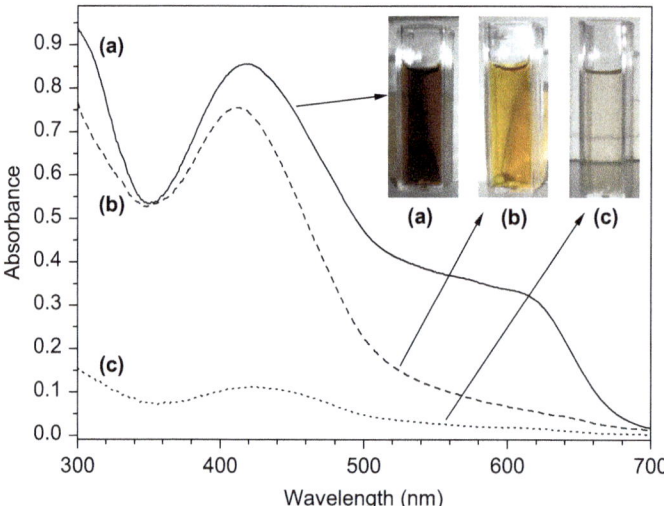

Figure 4.11   UV-visible spectra and images of a textile wastewater before and after biocatalysts contact: starting spectrum and solution (a), final spectrum and solution after treatment with free-SBP (b), final spectrum and solution after contact with M-SBP (c).

three dyes was diluted to a concentration suitable for spectrophotometric measurements. The tests were carried out following the modification of the UV-visible spectrum until no changes were observed. The results of these experiments are shown in Figure 4.11, together with the corresponding sample images.

The starting solution (Figure 4.11a) was coloured dark-brown and shows a very broad absorption band (approximately from 340 to 650 nm). By reaction with the free-SBP/H_2O_2 system, a fast decrease of the absorption at wavelength larger than 500 nm and a slight reduction of the absorption at lower wavelengths can be observed. The reaction reaches equilibrium in a couple of hours even in the presence of a low concentration of SBP. The wastewater solution becomes orange (Figure 4.11b). On the contrary, in the presence of a monolith functionalised with SBP, the reaction keeps going for 24 h and leads to an almost complete decoloration of the solution (Figure 4.11c). On the basis of more detailed experiments, not reported here for sake of brevity, we can affirm that this effect is due to the synergic action of enzyme and monolith: the dyes adsorbs on the silica surface, and degrades in the biocatalytic reaction.

The preliminary experiments carried out on these systems, and described above, indicate that their potential is very high.

The environmental impact of the monolith production is very low, because only water is employed as solvent and, moreover, because the use of a product derived from collection of green wastes favours and motivates towards separate waste collection.

Considering the applications of these devices the use of silica allows the surface of monoliths to be changed, modulating its properties *via* physical and/or chemical modification procedures easily available in the literature. Finally, a wide number of enzymes can be chosen and immobilised on silica materials, and this opens the way to potentially endless applications concerning, for instance, wastewater/soil bioremediation.

4.4 Food Wastes as Source of Products with Photochemical Properties for Environmental Applications

4.4.1 Photoinduced Generation of Reactive Species by HS

It has been demonstrated that the organic fraction of SBO has a similar chemical structure to humic substances (HS).[46] Therefore, it is expected that the photochemical properties of SBO are similar to those observed for HS. With this in mind, and considering that the photochemistry of HS has been widely studied, an overview of the photochemically mediated processes induced by HS is given below in order to gain insight into the possible applications of SBO as photosensitisers for wastewater treatment.

In natural waters, humic substances represent the main fraction of dissolved organic carbon that absorbs solar radiation and, therefore, they play an important role in aquatic photochemistry. Figure 4.12 shows the absorption spectra of water containing different kind of humic substances. In general, the absorption gradually decreases on increasing the wavelength. On the other hand, a simplified example of the solar spectrum is shown in Figure 4.12. It can be seen that most of the solar energy absorbed by HS occurs

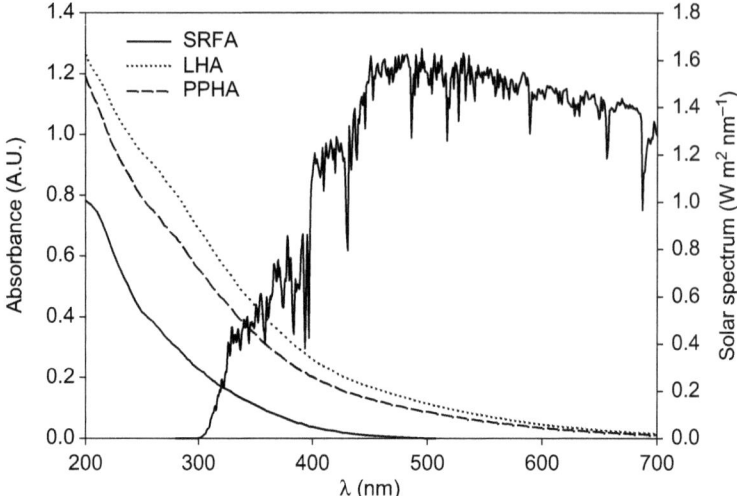

Figure 4.12   Solar spectrum and UV-vis absorption spectra of the different HS
(20 mg L^{-1}) in water at pH 7.0. SRFA = Suwannee River Fulvic acid;
LHA = Leonardite humic acid; PPHA = Pahokee peat humic acid.
Standard solar spectrum AM1.5 Global (ASTM G173-03).

between 300–700 nm. The primary excitation step therefore has to be between
95 and 41 kcal mol^{-1}. This energy is sufficient to initiate a number of different
photochemical processes.

The absorption of light by the HS chromophorous can lead to the formation
of several reactive species, including free radicals and hydrated electrons (e_{aq}^-),
reactive triplet states ($^3HS^*$), singlet oxygen ($^1O_2$), hydroxyl radicals (HO$^\bullet$),
superoxide ($O_2^{\bullet-}$) and hydrogen peroxide.[47–51] These reactive species may
undergo secondary thermal reactions that have important consequences with
respect to the modifications of pollutants, regulations of the redox properties of
natural waters, and the decompositions of HS.

Scheme 4.1 shows a simplified mechanism, which includes the possible
pathway that could lead to the formation of the reactive species.

The formation of the hydrated electron ($e_{(aq)}^-$) is thought to result from the
photoejection of an electron from the excited-state HS (Scheme 4.1). The quantum
yields for production of $e_{(aq)}^-$ is very low in natural conditions and it decreases
precipitously with increasing wavelengths. For Suwannee River fulvic acid, Thomas
et al. found that the quantum yields ranged from 1.9×10^{-4} at 296 nm to 7.9×10^{-6} at
366 nm.[52] The $e_{(aq)}^-$ is a highly reactive, strongly reducing species, which may react
through the direct reductive dehalogenation of organic pollutants, see eqn (4.1).
However, in oxygenated solutions, the dehalogenation is inhibited due to the $e_{(aq)}^-$
reacts predominantly with oxygen[53] ($k = 1.9 \times 10^{10}$ M^{-1} s^{-1}), see eqn (4.2).

$$e_{(aq)}^- + RHX \rightarrow {}^\bullet RH + X \tag{4.1}$$

$$e_{(aq)}^- + O_2 \rightarrow O_2^{\bullet-} \tag{4.2}$$

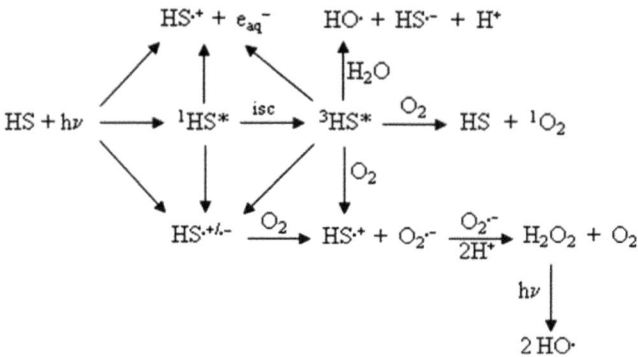

Scheme 4.1 Mechanisms of reactive species formation.

Singlet oxygen ($^1O_2$), the first excited state of molecular oxygen, is generated by energy transfers between the triplet excited state of HS ($^3HS^*$) and the ground-state oxygen (Scheme 4.1). This reaction has been shown to be pH-independent in the range 4.5–9.[54] The quantum yield of $^1O_2$ formation (Φ_Δ) depends on several parameters, mainly the irradiation wavelength and the nature of the HS. Literature values for Φ_Δ exciting between 300 and 400 nm generally range between 1 and 3%.[55,56] The $^1O_2$ formed is rapidly quenched by water[11] ($k_d = 2.3 \times 10^5$ s^{-1}) resulting in a low steady-state concentration. In turn, based on changing of the $^1O_2$ lifetime with HS concentration, several authors proposed that HS also act as effective quenchers of $^1O_2$.[57–59] On the other hand, $^1O_2$ is known to be a selective oxidant that reacts with electron-rich olefins, conjugated dienes, sulfides and phenols.[60,61] Basically, two mechanism are involved in the deactivation of $^1O_2$ with an organic compound; i) physical quenching, with no loss of the parent compound; ii) Chemical reaction, with the formation of a new product. The latter involves a different mechanism, such as 1,4- addition to conjugated dienes; oxidation of secondary alcohols to ketones and formation of organic hydroperoxides in the α-position to the double bond or aromatic ring.

It is accepted that the photogeneration of $O_2^{\bullet-}$ mainly arises from the population of charge-transfer states ($HS^{\bullet+/\bullet-}$) of the HS and subsequent reduction of oxygen to $O_2^{\bullet-}$ and from the oxidation of $^3HS^*$ by O_2[62] (Scheme 4.1). Here, $HS^{\bullet+/\bullet-}$ represents an intramolecular charge-transfer complex with diradical character. Alternative routes to $O_2^{\bullet-}$ formation could be the reduction O_2 by $e^-_{(aq)}$, see eqn (4.2) and the reaction of $^1O_2$ with HS, see eqn (4.3), but the relative contributions of these processes are unknown.

$$^1O_2 + HS \rightarrow O_2^{\bullet-} + HS^{\bullet+} \tag{4.3}$$

Superoxide is a relatively stable radical and a strong nucleophile in aprotic solvents,[63] but it is considered essentially unreactive toward oxidised organic compounds in deionised water, because it forms hydrogen bonds and undergoes rapid disproportionation.[64] On the other hand, similarly to the addition of aprotic solvents, the presence of solid surfaces also enhances the reactivity of superoxide in water with highly oxidised contaminants such as perchloroethylene and carbon tetrachloride.[65]

Hydrogen peroxide can act as either an oxidising or a reducing agent. H_2O_2 is a secondary product of light-induced reactions, however, it is relatively stable in water and may accumulate to concentrations that far exceed the other reactive species. The photochemical formation of H_2O_2 is thought to result from the disproportionation of superoxide, as shown in Scheme 4.1. Quantum efficiency values for H_2O_2 generation generally range between 0.02 and 0.5% at irradiation range 300–500 nm.[62,66,67] The H_2O_2 formed may be a source of free radicals and as such may be of major significance in indirect photolytic processes. Although H_2O_2 may not react directly with many organic pollutants, it may be important in metal speciation.[67]

Hydroxyl radical ($HO^{\bullet}$) is a highly reactive oxidant that unselectively reacts at near diffusion-controlled rates with most organic substrates. For this reason the formation of $HO^{\bullet}$ radical upon irradiation of HS has been widely studied.[68–71] However, there is much controversy regarding the mechanism of $HO^{\bullet}$ photogeneration induced by HS. There is evidence for several pathways, including oxygen-dependent, oxygen-independent and hydrogen-peroxide-dependent paths, but the mechanism still remains unclear.[69] Possible pathways are through the photolysis of hydrogen peroxide and through the abstraction of a hydrogen atom from water by the excited triplet state of certain substituted benzoquinones, which are components of HS (Scheme 4.1). In the presence of iron, other possible sources of $HO^{\bullet}$ radicals could be through Fenton's reactions. HS is known to form very stable complexes with iron. Upon irradiation, a redox reaction can occur, forming Fe(II) and free radicals from organic matter, see eqn (4.4). In neutral medium, Fe(II) is readily converted into Fe(III) by reaction with O_2, see eqn (4.5), but in acidic solutions, Fe(II) is probably oxidised by H_2O_2 formed and $HO^{\bullet}$ radicals are generated, see eqn (4.6).

$$HS - Fe(III) + h\nu \rightarrow Fe(II) + HS^{\bullet} \qquad (4.4)$$

$$Fe(II) + O_2 \rightarrow Fe(III) + O_2^{\bullet -} \qquad (4.5)$$

$$Fe(II) + H_2O_2 \rightarrow Fe(III) + HO^- + HO^{\bullet} \qquad (4.6)$$

The triplet state of HS ($^3HS^*$) is also photochemically a reactive molecule and direct reactions between $^3HS^*$ and organic substrates are likely to occur. They can be classified as energy, electron or hydrogen-atom-transfer reactions.[72] In energy-transfer (or photosensitation) reactions, $^3HS^*$ can transfer their energy to the ground state of the organic molecule:

$$^3HS^* + P \rightarrow HS + {^3P^*} \qquad (4.7)$$

This reaction is likely to occur if the energy level of the triplet state of the acceptor molecule is lower than that of HS. Few examples of these kinds of reactions have been reported in the literature (*e.g. cis*-isomerisation of 1, 3-pentadiene photoinduced by HS [73]). On the other hand, electron or hydrogen-atom transfer induced by $^3HS^*$ have been observed for several oxidisable organic molecule, such as aniline, phenolic compounds, amine drugs and

Phenylurea herbicides.[74–76] Recent works have shown that, among the reactive species formed upon irradiation of HS, ^{3}HS* play a major role in the degradation of some pharmaceutical products in waters.[77–79]

On the other hand, it has to be noted that most of the reactive species formed are able to react with HS. Thus, photodegradation of HS also occurs when HS are irradiated. Experiments carried out with chemical probes showed that the participation of $^1O_2$ and $HO^{\bullet}$ in the photodegradation of HS seems to be more relevant than $O_2^{\bullet -}$ and the effect of H_2O_2 was found to be negligible.[80]

4.4.2 Applicability of SBO in Wastewater Treatments

Among the possible uses of soluble bio-organic substances, it is worth exploring the applicability of these substances in wastewater treatment, based on their photophysical properties. The increasing demand of our society for water reaching the desired levels of quality makes it necessary for the development of alternative methods of wastewater treatment. In this context, the use of photochemical processes based in solar light might constitute a very convenient alternative from both the economical and ecological point of view.[81]

The contribution of sunlight to the self-depuration of aqueous systems is well known. Photochemical processes are among the major abiotic pathways for the removal of chemicals in natural waters.[82] Some pollutants are directly photolysed, as they are able to absorb in the UVA-visible range of the spectrum; however, in most cases, the reaction occurs through an indirect mechanism involving reactive species such as singlet oxygen, hydroxyl radical or a superoxide ion. Photochemical generation of these species can be mediated by different substances present in water, among them natural organic matter and in particular humic substances.[79] Although these processes have been demonstrated to be efficient in natural systems, where the pollutants concentrations are typically in the range of ng/L, they are not useful for the decontamination of effluents containing higher concentrations of chemicals as the generated amounts of reactive species are low and hence, reaction times would be too long to be implemented.

A possibility to enhance the efficiency of the solar process is the use of photocatalysts. These photocatalysts should not only absorb efficiently sunlight, but also be relatively cheap, environmentally friendly and able to work under different experimental conditions. Two different systems have been employed for this purpose, namely titanium dioxide and photo-Fenton. The first one is a stable nontoxic solid, that upon irradiation with wavelengths below 400 nm is able to generate highly reactive species such as hydroxyl radicals ($\bullet$OH) or holes in the valence band of the semiconductor that act as electron acceptors.[83] The limitations of this method are the separation of the photocatalyst after the treatment, or the low fraction of sunlight that can be employed (less than 10%). On the other hand, photo-Fenton consists in a combination of catalytic amounts of iron salts and sacrificial hydrogen peroxide (see eqns (4.6) and (4.8) for a simplified mechanism); this reagent has proven to be more efficient than TiO_2, although its major limitation is that highly acidic pHs (around 3) are required.[84] Nonetheless, this process is now at the stage to be tested on an industrial scale.

$$\text{Fe(OH)}^{2+} + h\nu \rightarrow \text{Fe}^{2+} + \bullet\text{OH} \qquad (4.8)$$

These processes have been implemented not only at laboratory scale, but there are a few preindustrial plants able to treat important volumes of effluent. Those plants are generally based on compound parabolic cylindrical collectors (CPCs). Those systems consists in two parabolic reflecting surfaces (generally made of aluminium) that are able to focus direct and disperse radiation in a glass tube through which the water flows,[85] as shown in Figure 4.13.

Alternatively, some organic dyes are known to be photochemically active and to generate, upon irradiation, highly reactive species. Different families of organic compounds have been checked for this purpose, namely acridines, anthracenes, anthaquinones, chlorines, porphyrines or phthalocyanines; other compounds such as pyrylium and thiapyrylium salts, methylene blue or rose bengal; all of them show important absorption bands in the UV-A or visible range of the spectrum, and hence, sunlight can be used as an irradiation source.[86]

The real applicability of these compounds is rather limited by their toxicity and/or low stability, and they might only be used as heterogeneous photo-catalysts, hosted onto different supports. However, they can be very useful for mechanistic purposes as the reactive species that they are able to generate are well defined or can be more easily determined than in more complex systems, as in the case of humic substances.

Figure 4.13   Pilot plant based on CPCs.

Different mechanistic pathways have been proposed for the reaction of a photocatalysts (P) with the substrates (s). One possibility is an electron-transfer process from the singlet or triplet excited state of the photocatalyst, as described in eqns (4.9)–(4.12).

$$P + h\nu \rightarrow {}^1P \tag{4.9}$$

$${}^1P \rightarrow {}^3P \tag{4.10}$$

$${}^1P + C \rightarrow P^- + C^+ \tag{4.11}$$

$${}^3P + C \rightarrow P^- + C^+ \tag{4.12}$$

This mechanism has been reported to play a very important role in the indirect photolysis of pollutants in the presence of humic substances, as will be discussed later in this chapter.

Another possibility is an electron-transfer process between the photocatalyst and water. This process is able to produce a hydroxyl radical ($\bullet$OH), which can be found among the strongest oxidising species:

$${}^1P + H_2O \rightarrow P^- + \bullet OH + H^+ \tag{4.13}$$

$${}^3P + H_2O \rightarrow P^- + \bullet OH + H^+ \tag{4.14}$$

Singlet oxygen ($^1O_2$) can also be generated; in this case, the mechanism involves an energy transfer process between oxygen and the triplet state of P formed according to eqns (4.9) and (4.10):

$${}^3P + O_2 \rightarrow P + \bullet^1O_2 \tag{4.15}$$

Finally, alternative processes in which a preassociation of the photocatalyst and the substrate plays an important role (eqns (4.16)–(4.18)), or formation of superoxide anion can also explain the reactivity of some compounds (eqn (4.19))

$$P + S \rightarrow PS \tag{4.16}$$

$$PS + h\nu \rightarrow PS^* \tag{4.17}$$

$$PS^* \rightarrow P^- + C^+ \tag{4.18}$$

$$P^- + O_2 \rightarrow P + O_2^- \tag{4.19}$$

With this background, different possible applications for SBO in wastewater processes, based on their photochemical properties, deserve to be investigated. First, it is interesting to determine the ability of these materials to degrade some pollutants. However, in addition to their performance, other issues should be determined, namely their optimal concentration, the stability and biocompatibility if they are employed in homogeneous catalysis and a possible strategy to host these compounds onto different supports to allow elimination from the solution after the treatment or recycle for further use.

In addition to this, SBO might be useful to apply a photo-Fenton process in neutral media. The use of complexing agents has been studied to prevent iron precipitation. In particular, humic acids have been employed for this purpose;[87]

thus, SBO might show similar behaviour. Furthermore, SBO are obtained from solid wastes and significant amounts of iron are present in their chemical composition, which might be useful to drive a mild photo-Fenton process.

4.4.3 Case Studies

The photosensitising properties of soluble bio-organic substances (SBO) obtained by urban biowastes have therefore been investigated for the degradation of many compounds of environmental interest.[88–91]

Photodegradation experiments have been carried out using mainly the two devices described below (Figures 4.14 and 4.15):

- Closed cylindrical Pyrex glass cells (4.0 cm diameter and 2.3 cm height), containing 5 mL of sample, irradiated by a 1500-W xenon lamp (Solarbox, CO.FO.MEGRA, Milan, Italy), equipped with a 340-nm cutoff filter. The intensity of the incident radiation, measured with a UV-Multimeter system, was $26.7\,W\,m^{-2}$. The samples were magnetically stirred during irradiation.
- A Pyrex stirred batch reactor (500 mL) equipped using a medium pressure 125 W mercury lamp (Helios-Italquarz, Milan). The lamp was surrounded by a Pyrex glass jacket in which circulated cooling water in order to keep the temperature of the solution inside the reactor constant (25 °C). The average intensity of the radiation, measured at 25 cm from the jacket and 8 cm lower from its top, was $6.8\,W\,m^{-2}$. Samples were irradiated in the reactor under continuous stirring and the sampling was performed at various reaction times by taking a fixed amount of solution from one of the three openings.

4.4.3.1 Dye Degradation

An intriguing motivation for studying SBO photosensitising effect for the degradation of azo-dyes is related to SBO surfactant properties and it is

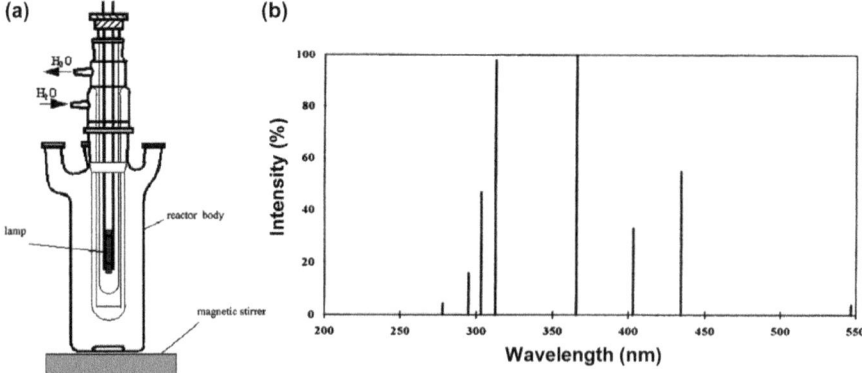

Figure 4.14 (a) Schematic representation of photochemical reactor. (b) Emission spectrum of a medium-pressure Hg lamp.

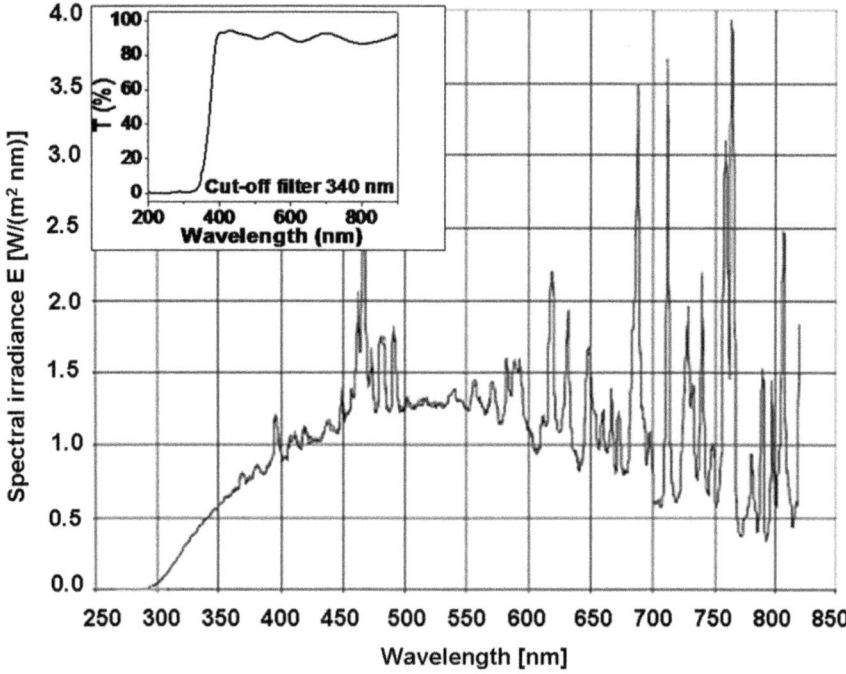

Figure 4.15 Emission spectrum of a 1500-W Xe lamp equipped with a soda-lime glass UV filter to simulate outdoor exposure. Inset: UV-vis transmittance (*T*%) spectrum of the 340-nm cutoff filter.

represented by the perspective to use SBO at two stages: in the dyeing process, *i.e.* to improve the dyeing process efficiency, and to help the removal of the residual dye from the exhaust dye bath. Photodegradation assisted by SBO has been investigated as an alternative solution for the remediation of industrial dye effluents, taking three typical commercial sulfonated azo-dyes, *i.e.* ethylorange (EO), Orange I (OI) and Orange II (OII), as probe molecules. These azo-dyes, in aqueous solution at 5 mg L^{-1} concentration have been irradiated in the presence of variable amounts of SBO.

Figure 4.16 clearly shows that an increase of the SBO/dye ratio, yields higher dye% abatement. Further experiments were performed to assess photodegradation kinetics for the three dyes by working at the optimum SBO concentration. The results reported in Figure 4.17 demonstrate that in all cases nearly quantitative dyes abatement can be achieved with kinetics following a pseudofirst-order law.

In order to give more insight into the dye degradation process, dye abatement rate, solution bleaching and sulfate evolution (as indicator of the dye mineralisation) have been examined and compared. Figure 4.18 reports the results obtained for Orange I; it can be noted that photobleaching is delayed compared to the dye disappearance that, could be due to the formation and accumulation of colored intermediates. The evolution of an amount of sulfate

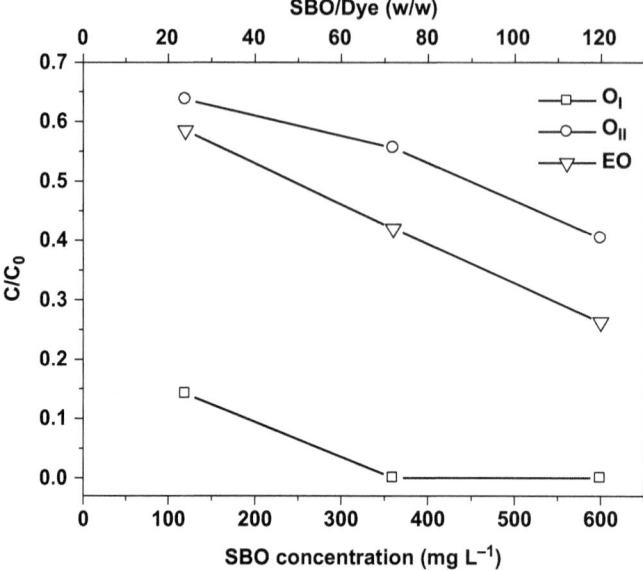

Figure 4.16 Dye C/C_0 ($C_0 = 5$ mg L^{-1}) upon three hours irradiation *versus* SBO mg L^{-1} concentration (lower abscissa axis) and SBO/Dye w/w ratio (upper abscissa axis) in solution. Experiments performed in a closed Pyrex® cell with a xenon (1500 W) lamp (Solarbox) and a cutoff filter for wavelengths below 340 nm.

lower than the stoichiometric value suggests that the intermediates are still bearing the sulfonic group. Analogous results were obtained for Orange II and ethylorange.

The irradiated solutions of the three azo-dyes were successively analysed by HPLC-MS in order to identify the intermediates formed during their photodegradation. Based on the observed intermediates, the most relevant dye degradation path was established to be: i) hydroxylation of the parent molecule to the aromatic ring; ii) N-dealkylation (EO); iii) azo-group reduction (OI and OII). Analogous degradation paths have been previously described in the case of the degradation of these dyes in the presence of TiO$_2$ suspension.[50] This aspect gives preliminary information on the possible photodegradation mechanism in the presence of SBO, that is the involvement of reactive oxygenated species (ROS), well established in the case of TiO$_2$-mediated Photocatalysis.[92]

4.4.3.2 Naphthalene Sulfonates Degradation

To extend SBO application to other substrate classes, naphthalene sulfonates have been also considered as probe molecules. Sulfonic acids are frequently found in aqueous industrial effluents. They contribute to the content of dissolved organic sulfur (DOS) in water,[93] a frequently used water quality control parameter. Within this class of compounds aromatic sulfonates

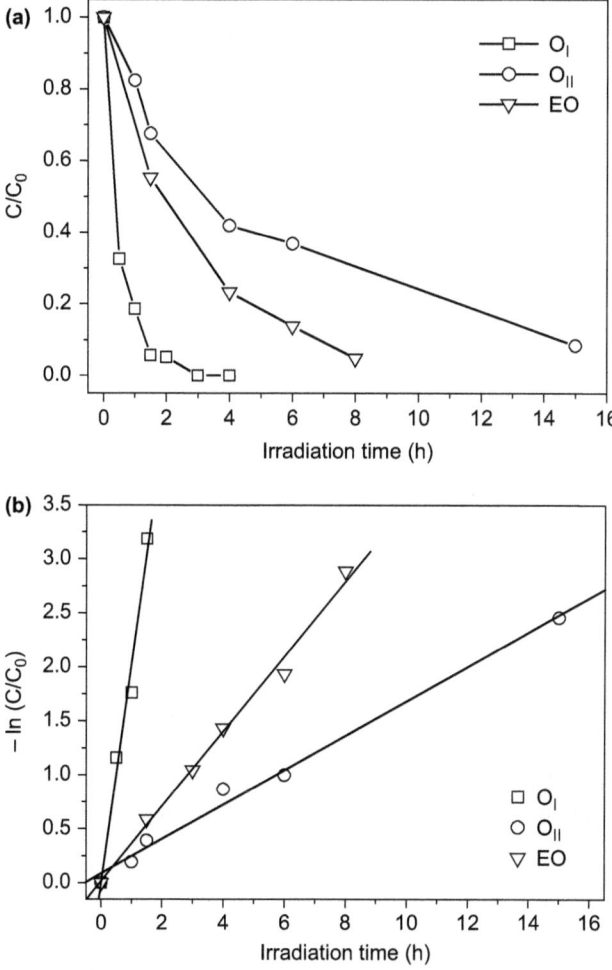

Figure 4.17 a) C/C_0 versus irradiation time for azo-dyes at $5\,\mathrm{mg\,L^{-1}}$ starting concentration in the presence of SBO at $360\,\mathrm{mg\,L^{-1}}$ concentration. b) $-\ln C/C_0$ versus irradiation time.

provide 40% DOS. Although the environmental impact of linear alkylbenzene sulfonates (LAS) has been widely investigated, little research has been dedicated to benzene and naphthalene sulfonates. In contrast to readily biodegradable LAS and benzene sulfonic acid, polycyclic aromatic sulfonates are reported to be rather persistent in wastewater effluents, surface waters and landfill leachates.[94,95]

Naphthalene sulfonates are widely employed in many industrial processes as dispersants, stabilisers, suspending and wetting agents, intermediates for dye synthesis. Even if these compounds are not significantly toxic, their degradation processes are worth investigating due to their widespread use, high stability and

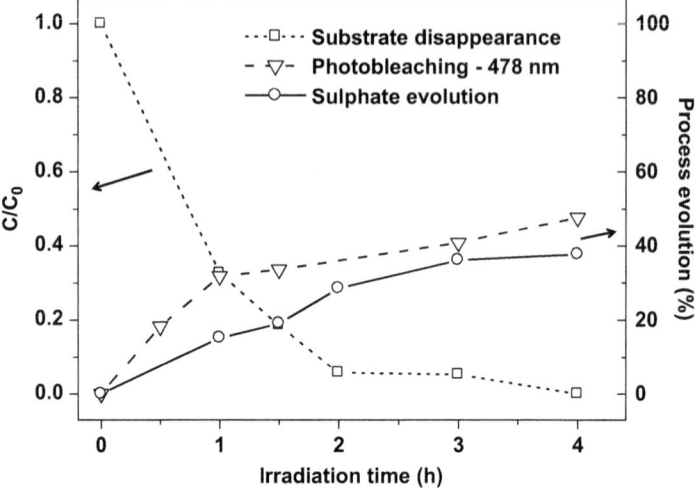

Figure 4.18   Degradation of Orange I at 5 mg L^{-1} initial concentration in the presence of SBO (360 mg L^{-1}) *versus* irradiation time: dye disappearance, photobleaching and sulfate evolution.

high solubility in water, which result in their transport and accumulation, even at considerable distances from the effluent source. Four different naphthalene sulfonates, 1-naphthalenesulfonic sodium salt (1-NS), 2-naphthalenesulfonic sodium salt (2-NS), 1,5-naphthalenedisulfonic acid (1,5-NdS) and 2,6-naphthalenedisulfonic disodium salt (2,6-NdS), have been used as probe molecules.

The specific SBO used in this study, referred to as AC8, was sourced from a UBW that was the product of the aerobic digestion of a 2:1 food/green residue mixture aged for 110 days.

Figure 4.19 shows the results when solutions containing 20 mg L^{-1} of the probe substrates were irradiated for 4 h in the presence of 50–500 mg L^{-1} AC8. As previously observed during the photodegradation of azo-dyes, for all the examined sulfonates, the measured degradation yield increases with the AC8 concentration, reaching a 15–30% plateau value at about 150–200 mg L^{-1} AC8.

Further work was carried out by irradiating 20 mg L^{-1} of the probe naphthalene sulfonate solutions in the presence of 150 mg L^{-1} AC8 to investigate the effect of the irradiation time. In this case, both the substrate and the sulfate concentration were monitored as indicators of the substrate degradation efficiency. The data, based on substrate concentration depletion in solution, indicate that after 24 h irradiation the substrate molar concentration decreased from about 6–8×10^{-5} to 2–1.5×10^{-5}, corresponding to 65–80% substrate removal. The photodegradation of the sulfonates appears to have been slightly greater than that of the disulfonates. The data showing sulfate concentration increases give additional information about the photodegradation process. It may be observed that for all the examined substrates, the sulfate release rate

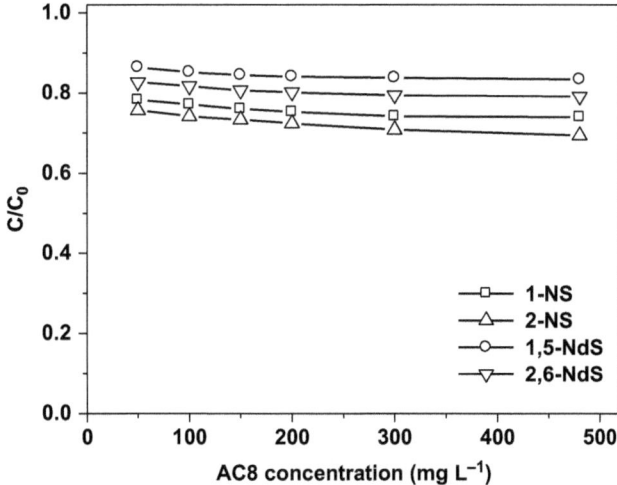

Figure 4.19 Substrate degradation *vs.* AC8 concentration after 4 h irradiation time. Initial substrate concentration $= 20$ mg L^{-1}.

in solution is much slower than the substrate depletion rate. After 24 h irradiation, only 15% and 30% of organic sulfur can be found in solution as sulfate for NS and NdS, respectively, compared to the higher percentage of depletion of the probe substrate (about 80% for NS and 70% for NdS). This evidence allows the hypothesis that both desulfonated and sulfonated intermediate products are formed during photodegradation of the probe substrates. The delayed release of inorganic sulfur does not necessarily hinder the complete mineralisation of the organic substrate. An analogous situation has in fact already been reported for the photodegradation of sulfonates in the presence of TiO_2,[96] in which complete mineralisation of organic carbon and sulfur was demonstrated to occur later than the complete abatement of the probe substrate.

4.4.3.3   4-Chlorophenol Degradation with Different SBO

Due to its environmental relevance, 4-chlorophenol was also investigated as a probe molecule, in order to compare possible different photosensitising effects when using a different SBO. Four different SBO were tested, all sourced from UBW sampled from the process lines of ACEA Pinerolese waste treatment plant in Pinerolo (Italy). The UBW were the digestate (FORSUD) recovered from the plant biogas production reactor fed with the organic humid fraction from a separate source collection of urban refuse, and three other materials obtained in the compost production section from different bioresidues: *i.e.* (i) CVT365 obtained from urban vegetable (V) residues aged for 365 days, (ii) CVD obtained from FORSUD and V mix aged for 110 days, and (iii) CVDF obtained from FORSUD, V and municipal sewage sludge (F) mix aged for 110 days. The UBW were processed further in a pilot plant[97] made available from Studio Chiono e Associati in Rivarolo Canavese, Italy.

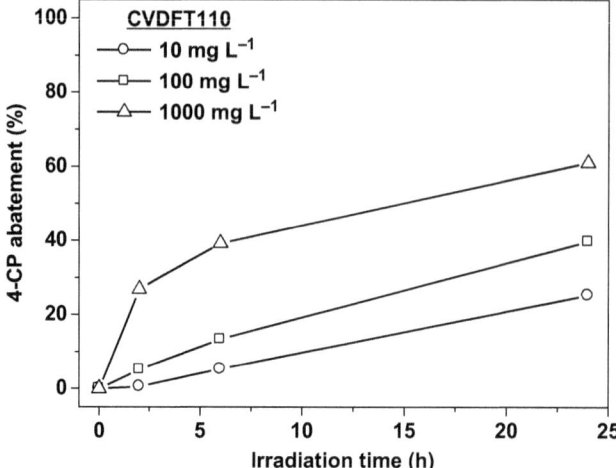

Figure 4.20   Effect of CVDFT110 concentration on the 4-CP $(10\,mg\,L^{-1})$ degradation.

Preliminary experiments performed on 4-CP aqueous solutions, irradiated up to 24 h in Solarbox in the absence of SBO showed that direct photolysis of 4-CP does not occur to an appreciable level.

In addition, because of the origin of the photosensitiser agent, it cannot be excluded *a priori* the presence of a bacterial component that could lead to partial degradation of the substrate even in the absence of photoactivation. The samples (4-CP $10\,mg\,L^{-1}$ and SBO $500\,mg\,L^{-1}$) were therefore analysed after being kept in the dark for 24 h; the results do not show any variation in the substrate concentrations.

The irradiation of 4-CP $(10\,mg\,L^{-1})$ was then carried out in Solarbox in the presence of the four SBO separately at different concentrations in the range from 10 to $1000\,mg\,L^{-1}$; Figure 4.20 shows the results obtained in the presence of CVDFT110; analogous behaviour was observed also when using the other SBO.

As can be seen, the presence of SBO favored the substrate degradation and a beneficial effect can be noted when SBO concentration increases from $10\,mg\,L^{-1}$ up to $1000\,mg\,L^{-1}$. In Figure 4.21 the four SBO are compared, by considering the 4-CP sensitised degradation after 24 h of irradiation. The observed trend suggests for each SBO an upper concentration limit, above which no further beneficial effects can be obtained.

4.4.3.4   Phenol and 4-Methylphenol Degradation with CVT230

Based on the encouraging results obtained with 4-CP degradation, other phenolic compounds have been considered for their photodegradation in the presence of SBO, namely phenol and 4-methylphenol (Ph and 4MP, respectively).

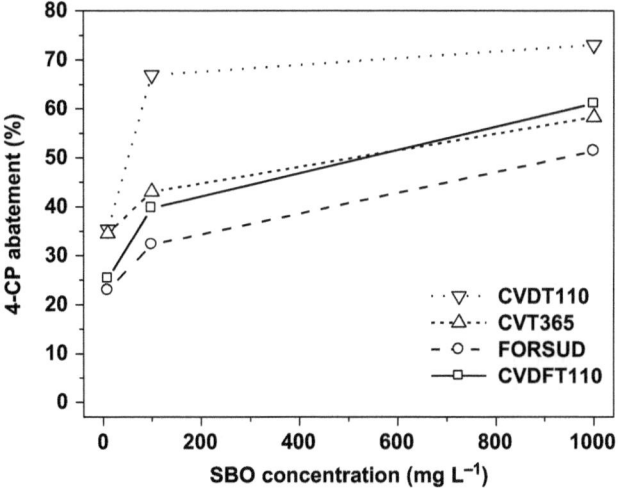

Figure 4.21 Percentage of 4-CP degradation in the presence of different amount of SBO, after 24 h of irradiation.

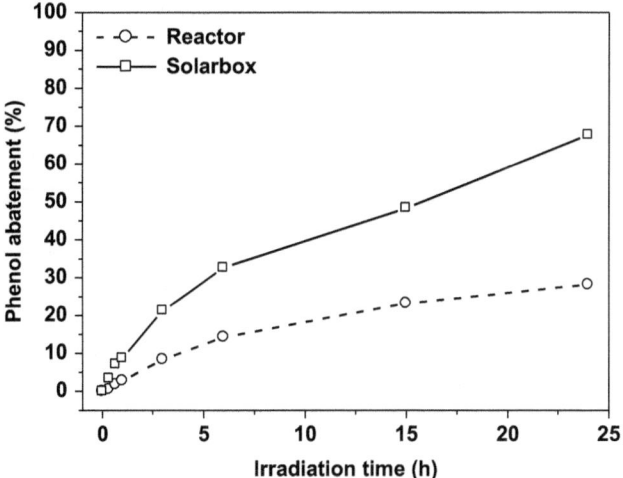

Figure 4.22 Phenol (1.0×10^{-4} M) degradation in the presence of CVT230 (500 mg L^{-1}). Phenol abatement vs. irradiation time - Solarbox and reactor.

For these substrates (1.0×10^{-4} M), experiments have been performed in the presence of the SBO CVT230 (500 mg L^{-1}), isolated from a green residue composted for 230 days.

The irradiations were performed in two different systems: Solarbox (xenon lamp with cutoff filter at 340 nm) and Reactor (medium-pressure Hg lamp).

In Figures 4.22 and 4.23 the kinetic profiles obtained for the abatement of the two substrates are reported. A faster degradation can be observed for 4-MP

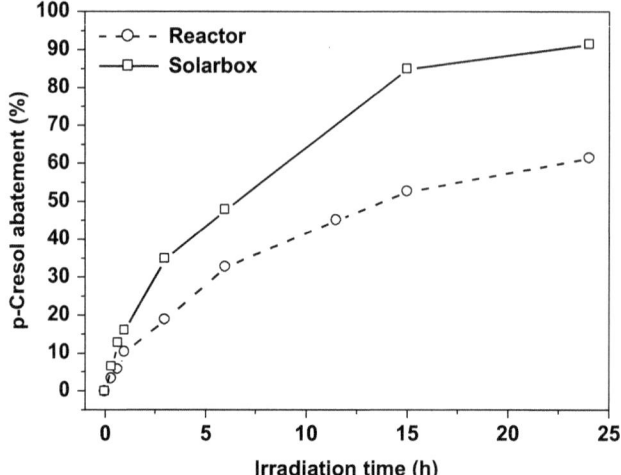

Figure 4.23 p-Cresol $(1.0 \times 10^{-4}\,\mathrm{M})$ degradation in the presence of CVT230 $(500\,\mathrm{mg\,L^{-1}})$. p-Cresol abatement *vs.* irradiation time – Solarbox and reactor.

Table 4.3 Kinetic constant calculated from the slope of the plot reported in Figures 4.22 and 4.23.

SOLARBO4	*Linear Fit*	r^2	$k_{obs}\ (h^{-1})$
Phenol	$y = 0.0939x + 0.0028$	0.9885	0.094
p-Cresol	$y = 0.1893x + 0.0066$	0.9919	0.189
REACTOR	*Linear Fit*	r^2	$k_{obs}\ (h^{-1})$
Phenol	$y = 0.0291x - 0.0018$	0.9872	0.040
p-Cresol	$y = 0.1125x - 0.0020$	0.9918	0.113

and both substrates reasonably fit a pseudofirst-order kinetic model, thus allowing the observed kinetic constant (k_{obs}) reported in Table 4.3 to be calculated.

Both in Solarbox and in the reactor the same reactivity order was found, *i.e.* 4-MP > phenol; moreover, it is evident that for the Solarbox, the degradation is much faster than in Reactor. In Solarbox, after 24 h irradiation, almost complete abatement of 4-MP, and 65% of degradation was observed for phenol; in Reactor 65% was measured for 4-MP and only 30% for phenol.

The explanation of the different performances of the two irradiation systems is not so straightforward, but some aspects possibly influencing the kinetics can be underlined:

- different light sources with different emission spectrum and radiation intensity;
- different geometry, and different thickness of the irradiated solution layer;
- different air volume/ solution volume ratio (much lower in the reactor).

This last aspect could affect the level of oxygen available to participate in the photodegradation process; previous studies have shown a positive effect of the oxygen both in the presence of the natural humic acids and of SBO.[89,98]

The effect of the oxygen on the photodegradative process was thus more heavily investigated in the successive experiments.

Different reaction environments were investigated by performing irradiation experiments in the Reactor under continuous gas bubbling (nitrogen or air) from one of the openings. No appreciable differences have been observed between the processes run in the closed system and that run under nitrogen. On the contrary, in the presence of oxygen the process becomes faster, and this fact suggests that the mechanism proposed in the literature for natural humic acids (HA), involving the reactive oxygenated species (ROS), is operating also for the SBO.

4.4.4 EPR Spin-Trapping Technique for the Hydroxyl Radical and Singlet Oxygen Determination

The mechanism responsible for the photosensitising properties of SBO has been investigated. Since previous studies reported reactive oxygenated species as main actors when humic acids are involved, rather than excited triplet states, supposed to be relevant in the presence of fulvic acids, the attention was focused on $^{\bullet}OH$ and the $^1O_2$ production when SBO are irradiated. Actually, SBO can be compared to humic rather than to fulvic acids, because of their estimated molecular weight and their stability in solution at pH values above the neutrality.

In the literature concerning photoassisted degradation processes occurring in water in the presence of natural organic matter, the study of $^{\bullet}OH$ and $^1O_2$ formation and evolution has been carried out mainly by using probes/scavengers or by electronic paramagnetic resonance (EPR) spectroscopy.

In the first case, suitable molecules react selectively with the species under investigation, yielding to a specific reaction product (in the case of probes) or inhibiting the kinetics of the photodegradation process of a target organic molecule (in the case of scavengers). In this work two type of scavengers were employed: 2-propanol as $^{\bullet}OH$ scavenger[99] and sodium azide as $^1O_2$ scavenger.[100]

In the second case, the monitoring of reactive species is performed by means of EPR using the spin-trapping approach.[101,102] The spin-trapping technique is employed in the detection and identification of short-lived free radicals. Spin trapping involves the addition of transient radical species to a diamagnetic compound (spin trap), with the formation of a paramagnetic persistent radical (spin-adduct), that can be detected by EPR spectroscopy.

The spin trap, generally a nitrone or nitroso compound, must be stable in the reaction conditions, and not take part in secondary reactions and undergo radical addition quickly in order to avoid further reactions of the radicals before the trapping.

Both approaches suffer from possible drawbacks of indirect methods; scavengers are not usually specific for a single radical and spin-adduct formation reactions may be affected by low yield and/or insufficient product stability.

The photosensitising properties of one SBO, namely AC8 was studied using 4-chlorophenol (4-CP) as the probe molecule.[103] First, the production of •OH and $^1O_2$ was monitored by EPR spectroscopy. Afterwards, the correlation between radical species evolution and photodegradation of 4-CP was investigated. Finally, the effect of $^1O_2$ and •OH scavengers on the 4-CP degradation process was analysed. 4-oxo TMP (45 mM) and DMPO (17.4 mM) were employed as trapping agents for $^1O_2$ and •OH, respectively. A preliminary optimisation of instrumental and experimental parameters was performed according to the available literature. All experiments were carried out by adding the spin trap in the cell before irradiating the SBO, with the EPR spectra acquired immediately after irradiation. Other preliminary studies were performed in order to obtain the maximisation of EPR signals and to reduce the signal/noise ratio.

4.4.4.1. *Singlet Oxygen*

TMP (45 mM) was employed as trapping agent for $^1O_2$ study, based on the reaction reported in Figure 4.24.

In order to assess the SBO concentration effect on $^1O_2$ production, experiments were carried out irradiating solutions containing AC8 in a concentration range from 0 to 2000 mg L^{-1}. These experiments were performed four times in order to verify the reproducibility; moreover, the limit of detection of the method, calculated as the average signal of the background plus three times its standard deviation that corresponds to the lowest examined concentration of 100 mg L^{-1}.

The trend of the signal intensity reported in Figure 4.25 evidences that $^1O_2$ production is directly proportional to AC8 concentration.

4.4.4.2 *Hydroxyl Radical*

DMPO was employed as a trapping agent for •OH study. As reported in Figure 4.26 the addition to DMPO of different radical species leads to different specific EPR spectrum.[104]

The typical EPR spectrum of the DMPO-OH spin adduct is reported in Figure 4.26. The spectrum is a 1:2:2:1 quartet, with a 14.8 G hyperfine splitting constant.

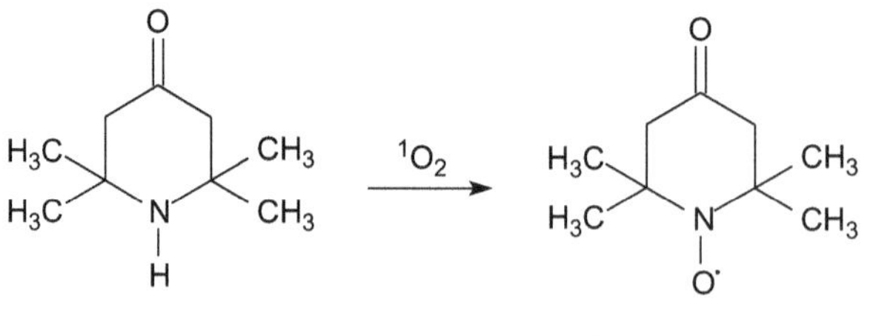

Figure 4.24   Formation of spin-adduct 4-oxo-TMP-$^1O_2$ (4-oxo-TEMPO).

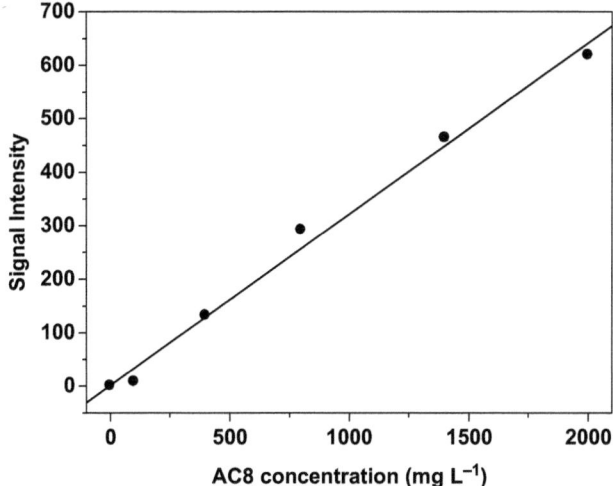

Figure 4.25 Intensity of 4-oxo-TEMPO EPR signal, after background subtraction, *vs.* AC8 concentration. 4-oxo-TMP concentration = 45 mM; irradiation time = 15 min. Linear fit of data, slope = 0.3198, intercept = 1.7913, $r^2 = 0,9920$.
Reproduced with permission from Wiley-VCH Verlag GmbH & Co. KGaA, Weinheim.

The effect of SBO concentration on $^{\bullet}$OH radicals production was investigated by varying the AC8 concentration in the range from 0 to 2000 mg L^{-1}.

Figure 4.27 shows the intensity of the EPR spectrum of the DMPO–OH adduct *versus* AC8 concentration. In contrast to 4-oxo-TEMPO the DMPO–OH spectrum intensity increases as SBO concentration increases until a maximum value about 20–50 mg L^{-1}, and then decreases at higher SBO concentrations.

This profile can be explained by hypothesising the simultaneous occurrence of two different processes: production and scavenging of $^{\bullet}$OH by irradiated SBO.

Photodegradation of SBO is proven by UV-Vis spectroscopic measurements. The SBO degradation process may well contribute to the decrease of hydroxyl radicals available for the DMPO–OH adduct formation, and thus cause the behaviour observed in Figure 4.27. The scavenging process becomes dominant at high SBO concentration.

4.4.4.3 Effect of $^{1}O_2$ and $^{\bullet}OH$ Scavengers on the 4-Chlorophenol Photodegradation

4-Chlorophenol (4-CP) was chosen as a model compound to investigate the existence of a correlation between the kinetics of its photodegradation, the $^{1}O_2$ and $^{\bullet}$OH production, and the addition of $^{1}O_2$ and $^{\bullet}$OH radical scavengers.

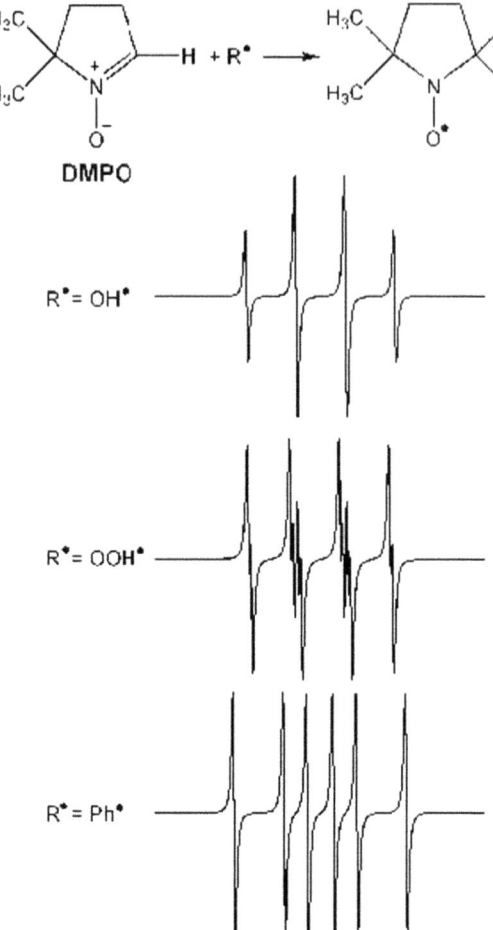

Figure 4.26 EPR spectra of DMPO with three different radical species.[104]

The substrate (5 mg L^{-1}) was irradiated in the presence of AC8 at two different concentrations: 100 mg L^{-1}, where the low $^1O_2$ production and the high $^{\bullet}$OH production are suggested from the data previously reported, and 1200 mg L^{-1}, whereby the ratio between $^1O_2$ and $^{\bullet}$OH production is greatly increased. Figures 4.28 and 4.29 show that the 4-CP degradation depends on the irradiation time, achieving 65 and 75% after 24 hours in the case of 100 mg L^{-1} and 1200 mg L^{-1} of AC8, respectively. In the same figures the effect of the addition to the AC8/4-CP solution of two different scavengers: 2-Propanol (0.01 M) as $^{\bullet}$OH scavenger and sodium azide (0.03 M) as $^1O_2$ scavenger is also shown.

At 100 mg L^{-1} AC8 concentration, in the absence of scavengers, after 24 h irradiation about 65% of 4-CP is degraded (Figure 4.28). Based on the relatively high $^{\bullet}$OH production and low $^1O_2$ production it could be hypothesised that $^{\bullet}$OH is mainly responsible for the 4-CP degradation.

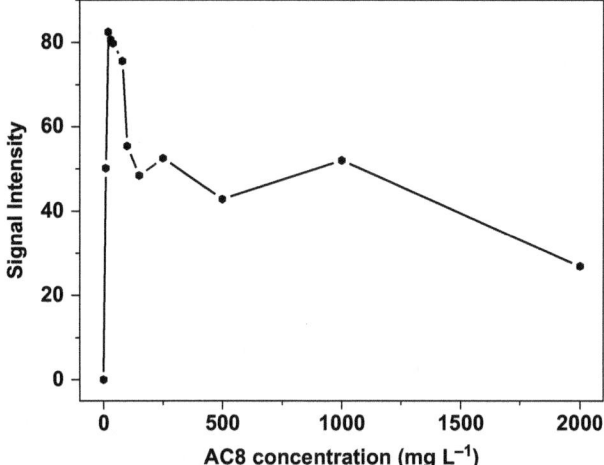

Figure 4.27 Intensity of DMPO-OH EPR signal vs. AC8 concentration. DMPO concentration = 17.4 mM; irradiation time = 3 min.
Reproduced with permission from Wiley-VCH Verlag GmbH & Co. KGaA, Weinheim.

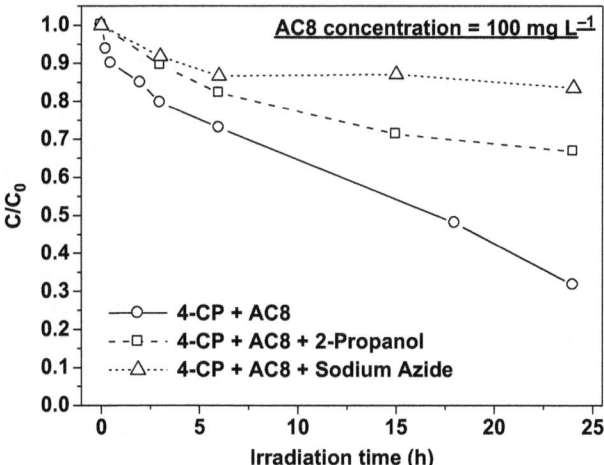

Figure 4.28 Degradation of 4-CP ($5\,mg\,L^{-1}$) in the presence of AC8 ($100\,mg\,L^{-1}$), 2-Propanol (0.01 M) and sodium azide (0.03 M).

The addition of 2-propanol, a •OH scavenger, led to significant reduction of the 4-CP abatement from 65% to 33%, in agreement with the previous hypothesis. On the other hand, in the presence of sodium azide, a $^{1}O_2$ scavenger, the 4-CP abatement decreased from 65% to 15%. Despite sodium azide is mainly reported as a highly selective $^{1}O_2$ scavenger, it is known to react also with OH radicals. This result could be explained if both •OH and $^{1}O_2$ play

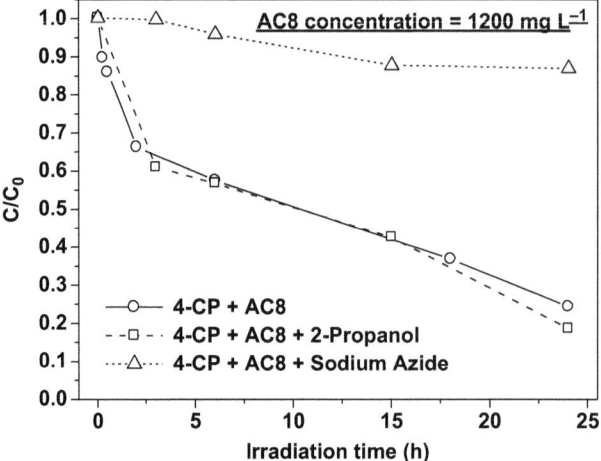

Figure 4.29   Degradation of 4-CP ($5\,mg\,L^{-1}$) in the presence of AC8 ($1200\,mg\,L^{-1}$), 2-Propanol (0.01 M) and sodium azide (0.03 M).

an active role in the 4-CP degradation, and the higher effect of sodium azide could thus be explained on the basis of its additional $^{\bullet}OH$ scavenging property.

Figure 4.29 shows that 24 h irradiation of 4-CP in the presence of $1200\,mg\,L^{-1}$ AC8 concentration and in the absence of scavengers is slightly more efficient than that at the lower AC8 concentration, yielding a 75% abatement. In agreement with the results obtained with $100\,mg\,L^{-1}$ of AC8, the 4-CP degradation could be reasonable due to both $^{\bullet}OH$ and $^{1}O_2$. However, as can be observed, only sodium azide significantly inhibited the photodegradation process, whereas the addition of 2-propanol seemed to have no appreciable effect at this concentration. The sodium azide effect seems consistent with the high $^{1}O_2$ production, about 40 times higher than that observed at $100\,mg\,L^{-1}$ AC8 concentration. The negligible effect of 2-propanol at $1200\,mg\,L^{-1}$ AC8 concentration may suggest that, in this case, $^{\bullet}OH$ does not play a significant role in 4-CP degradation.

These results suggest that the role played by $^{\bullet}OH$ radicals and singlet oxygen in 4-CP degradation is strictly related to their relative production and depends on the SBO concentration. This study, focused on $^{\bullet}OH$ and $^{1}O_2$, does not exclude the possible contribution of other reactive species, such as excited triplet states $(^{3}HS^{*})^2$, solvated electron (e^-), anion superoxide ($O_2^{\bullet-}$), hydrogen peroxide (H_2O_2) and hydroperoxyl radical ($HOO^{\bullet}$).

4.4.5   SBO Photodegradation

Due to its rather complex nature, the degradation of AC8 could not be evaluated by direct concentration measurements; however, indirect evidence can be obtained by UV-VIS spectroscopic measurements.

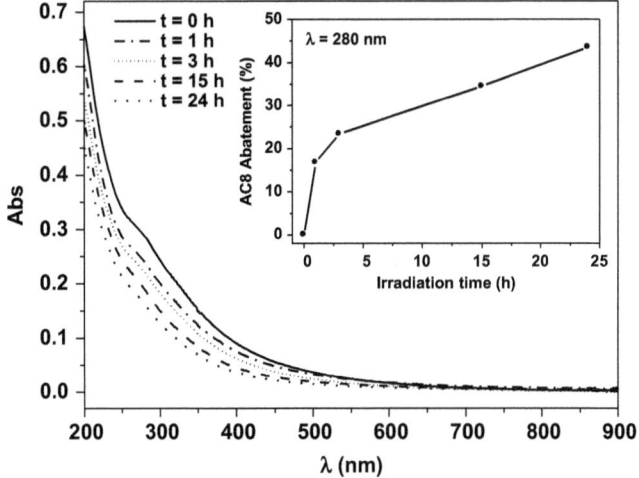

Figure 4.30 Absorbance (Abs) as a function of wavelength (λ) in UV-VIS spectra of 150 mg L^{-1} AC8 solution (recorded after diluting the samples by a factor of 10) at different irradiation times (t). Inset: relative Abs abatement at 280 nm vs. irradiation time.
Reproduced with permission from Elsevier.

All the examined SBO show a progressive abatement of their UV-vis spectrum upon irradiation (see Figure 4.30 as an example). These aspects introduce a role of competitor of SBO for the reactive species produced by SBO itself. The balance between reactive species production rate and their consumption by either an added organic substrate or by SBO itself, could in principle explain the plateau observed when the degradation rate is plotted *vs.* SBO concentration.

4.4.6 Toxicity Tests

When a degradation process aims to be technologically upgraded for its application to the treatment of real wastewaters, it is of fundamental importance to ascertain not only the disappearance of the pollutants, but also their degradation pathway. In fact, the evolution of a pollutant could result in intermediates or byproducts even more toxic than the starting material. In the classical AOP, the substrate mineralisation is easily monitored by analysing the total organic carbon (TOC) evolution. Moreover, detailed intermediates analysis and identification has become common, mainly by means of sophisticated chromatographic techniques. In the presence of SBO this traditional approach cannot be easily adopted; because of its high concentration, SBO contribute to the organic carbon balance as main constituent of the system and even an almost negligible SBO mineralisation influences the carbon balance. It becomes therefore very difficult to distinguish between the carbon decrease due to SBO mineralisation and the carbon decrease due to pollutant mineralisation.

Table 4.4 Biotoxicity data for different investigated substrates.

Investigated solution	Substrate abatement (%)	Inhibition Effect%
AC8[90]	0	4.6
AC8[90]	24 h	9.4
AC8/EO[105]	0	−5
AC8/EO[105]	95.2	5
AC8/O$_I$[105]	0	−12
AC8/O$_I$[105]	100	18
AC8/O$_{II}$[105]	0	18
AC8/O$_{II}$[105]	91.6	10
AC8/1-NS[90]	0	27
AC8/1-NS[90]	84.4	17
AC8/2-NS[90]	0	29
AC8/2-NS[90]	83.6	18
AC8/1,5-NdS[90]	0	18
AC8/1,5-NdS[90]	74.3	−33
AC8/2,6-NdS[90]	0	17
AC8/2,6-NdS[90]	67.8	41

Also, the intermediates study is not so straightforward due to the possible matrix effect, due to the SBO aggregate or their intermediates. In such a complex system, a valuable alternative could be a toxicological essay and in the last years an emerging assay is the acute toxicity monitoring by means of the Vibrio Fischeri luminescence inhibition. This test is a powerful tool for screening the toxic properties of a set of samples containing a multitude of chemical compounds. It allows a preliminary classification to be obtained of the overall toxicity based on the% luminescence inhibition effect. Table 4.4 reports the results obtained in the previously presented degradation tests.

The experimental data proves that BOS are efficient sensitisers in the photolysis of azo-dyes. The observed kinetic order and the type of intermediates generated by the dyes photolysis, assisted by BOS are similar to those reported in the photolysis of the same dyes assisted by TiO_2. This fact may suggest a similar mechanism in both cases, and warrants a specific investigation on the photodegradation mechanism induced by BOS.

4.5   Food Wastes as Source of Products for Agriculture

Different composts and the digestate from a biogas plant, and their respective alkali-soluble (SBO) and -insoluble (IOR) fractions have been tested as organic fertilisers in growth media.[106] The starting biowastes were obtained from the Acea Pinerolese waste-treatment plant located in Pinerolo (TO), Italy. The Acea plant contains four sections; two for the anaerobic (AN) and aerobic (AE) digestion of urban solid biowastes (UBW) from separate source collections, the third one for treating urban wastewaters (WWT) and the last one being a landfill area equipped for biogas collection (LBG). The four plant sections are interconnected to maximise biogas and compost yields from biowaste, thus minimising biorefuse disposal to landfill. In essence, the bio-organic (humid) fraction of solid urban waste entering the AN process is fermented to yield biogas and a solid digestate (FORSUD) containing residual organic matter not converted to biogas. The FORSUD material is mixed with home gardening and park trimmings residues (V) and/or with sewage sludge (F) coming from the WWT process. The bioresidues mix is then allowed to undergo fermentation under AE conditions to yield compost. The plant allows large operational flexibility to produce different types of compost depending on the nature and relative ratios of the bioresidues constituting the AE section feed: CV being V composted for 365 days, respectively; CVDF being a 35/55/10 w/w/w FORSUD/V/F mix composted for 110 days; CVD being a 33/67 w/w FORSUD/V mix composted for 110 days; CVF being a 35/10 w/w mix composted for 110 days. The biowastes were further processed in a pilot plant made available by Studio Chiono e Associati in Rivarolo Canavese, Italy,[97] each yielding the corresponding SBO and IOR fractions. Germination bioassay and plant growth experiments in a farm located in Piemonte (Italy) were performed. Seeds of cress (Lepidium sativum), white mustard (Sinapsis alba) and radish (Raphanus sativus) were used in the bioassay trials. Tagetes patula (marigold), Raphanus sativus (radish) and Spinacia oleracea were used in pot experiments performed under a hail screen or in a green house. Plants grown on commercial substrate, used as control, were compared with the same plants grown on the test medium obtained by mixing the above biowastes, SBO and IOR materials with the commercial substrate on a 30% (V/V) basis. The results showed that most materials did not promote the germination of the seeds. In the pot experiments, the dimension and dry weight of the plants grown on the commercial substrate mixed with the starting biowastes or the corresponding SBO were similar to those observed on the neat commercial substrate. Most of the SBO provoked the increase of the dry weight of the roots of the radishes, therefore an increase of the quality. The composts' IOR fractions also allowed growth of the plants similar to that obtained with the commercial substrate. In contrast, the FORSUD IOR was phytotoxic.

Sortino et al.[107] investigated CVD, CVD SBO and CVD IOR as organic fertilisers for tomato greenhouse cultivation in a farm located in south east Sicily (Italy). The products were applied to the soil at the rate of 1.1–1.2 ton ha^{-1} of organic matter. A commercial product, normally used in the farm's soil treatment, was also used for comparison. This product was declared by the supplier to contain meat, bone and animal blood meal. No significant differences were found in the soils physicochemical features due to the different treatments. However, the plant biometric data and productivity were significantly affected. Compared to the control and all other treatments, the CVD SBO exhibited the best performance by enhancing leaf chlorophyll content, improving plant growth and fruit ripening rate and yield. Solubility could be one reason for the superior performance of SBO as a tomato growth promoter. Particularly interesting is the appeared enhancement of leaf chlorophyll content in connection with the photosensitising properties of SBO, demonstrated in Section 4.4. The results led the authors to propose that SBO, under solar irradiation, might promote photosynthesis or organic carbon mineralisation depending upon the experimental conditions. In a successive study, Sortino *et al.*[108] carried out the cultivation of red pepper using the same soil and SBO as in the tomato cultivation experimental plan. However, for red pepper cultivation CVD SBO was applied to the soil at 0–700 dry matter kg ha^{-1} doses, corresponding to 0-0.5 organic matter ton ha^{-1}. The reported results show that the SBO applied doses did not yield detectable soil chemical composition changes relatively to the control soil. However, the plant leaf chlorophyll content, growth and productivity increased to maximum values upon increasing the soil treatment dose up to 35–140 dry matter kg ha^{-1}. The most remarkable results were the maximum productivity increases observed for the 140 dry matter kg ha^{-1} treatment dose compared to the control soil. The increases amounted to 90% for the precocious crop yield, to 66% for the total crop production and to 17% for the per fruit weight. The discovery that the highest effects occur at such low treatment dose suggests using the above soluble substances to enhance plant growth and productivity, while minimising the potential environmental impact of conventional fertilisers. Particularly interesting was the fact that the leaf chlorophyll content reached its peak level for the plants grown in the soil treated with 140 SBO dry matter kg ha^{-1}, while decreasing for higher SBO doses. The SBO dose–leaf chlorophyll content pattern was somewhat similar to the dose–effect relationship reported for SBO tested as photosensitiser for the abatement of organic pollutants in solution:[90] *i.e.* the rate of the probe molecule photodegradation increases upon increasing the soluble substances' concentration to a maximum value; higher concentrations cause no enhancement of the probe molecule photodegradation rate.

Certainly, the similarity of the dose–effect relationship pattern observed for the leaf chlorophyll with that observed for the photodegradation of organic pollutants, performed with the same soluble substances in the absence of any possible biochemical reaction,[90] may suggest the hypothesis that these substances performed a photosensitising activity also in photosynthesis. Understanding the mode of action of the soluble substances under the experimental conditions reported for the photodegradation of organic pollutants is simpler than in the case of red pepper cultivation. In the second case, the soluble substances must undergo a number of mass-transfer processes from the soil to the plant in order to perform their action. In these processes, a high number of chemical and biochemical factors are involved. The available data do not allow the hypothesis of the role of the soluble substances as photosensitisers for promoting photosynthesis to be confirmed. However, proving that biorefuse sourced SBO may enter the natural C cycle and be used to enhance solar-light efficiency for either C fixation or C mineralisation would be a fascinating research scope.

4.6   Perspectives to Develop a Biorefinery Fed with Biowastes

The results reported in this chapter suggest that urban food wastes mixed with other biowastes may be processed to obtain a wide range of products for use in the chemical industry, in material science, in the remediation of industrial effluents environmental and in agriculture. Further uses are likely to be demonstrated in the future. In principle, this situation may depict an intriguing scenario where urban wastes treatment plants may be turned into a biorefinery for the production of energy, chemicals and consumer's products with an environmentally friendly impact. A recent study[7] has evaluated the expected economical and environmental impacts deriving from a virtual scenario that comprises integration of SBO production technology into a waste-management plant treating urban biowastes by aerobic and anaerobic microbial digestion according to the current technology. The results point out that the integrated plant selling biogas and SBO is likely to yield six times higher earnings than the current plant whose revenue comes from biogas and compost. To make this scenario real, a major critical point to overcome is the risk connected to the market desirability of the SBO products. While biogas sales for energy production are guaranteed by the market demand and promoted by government policies through fiscal incentives, the refuse-derived biobased chemicals must face the competition of assessed synthetic chemicals and the market diversified and sophisticated demands. The second critical point is that such a biorefinery can only be developed from an existing waste-management plant that already performs the collection and separation of wastes. A waste-management plant, however, lacks the expertise in marketing chemicals, and therefore needs partnership to enter and establish the market for such products. One alternative is that the waste-management plant could use the SBO products for internal needs. Research in this direction might create real perspectives for starting a biorefinery fed with biowastes.

Acknowledgements

The author's are grateful to 7thFP IRSES-2010-269128-EnvironBos Marie Curie Action.

References

1. J. Kaminsky, (2004). Development of Strategies for Deployment of Biomass Resources in the Production of Biomass Power. NREL/SR-510-33524, available electronically at http://www.osti.gov/bridge.
2. P. Lusk, (1998). Methane Recovery from Animal Manures. The Current Opportunities Casebook. NREL/SR-580-25145, available at http://www.doe.gov/bridge/home.html.
3. W. E. Mabee, D. J. Gregg, C. Arato, A. Berlin, R. Bura, N. Gilkes, O. Mirochnik, 4. Pan, E. K. Pye and J. N. Saddler, *Appl Biochem Biotechnol.*, 2006, **55–7**, 129.
4. R. Hamidimotlagh, I. Nahvi, G. Emtiazi and S. Abedinifar, *Afr. J. Biotechnol.*, 2007, **6**, 1110.
5. C. Arato, E. K. Pye and G. Gjennestad, *Appl. Biochem. Biotechnol.*, 2005, **121**, 871.

6. J. Clark, *J Chem Technol Biotechnol*, 2007, **82**, 603.

7. E. Montoneri, D. Mainero, V. Boffa, D. G. Perrone and C. Montoneri, *Int. J. Global Environmental Issues*, 2011, **11**, 170.

8. E. Montoneri and C. Montoneri, Biopolimeri isolati da biomasse residuali e da fossili, relativi procedimenti di produzione ed usi. Università di Torino, patent number TO2011A001192, 22 dicembre 2011.

9. E. Montoneri, V. Boffa, P. Savarino, D. G. Perrone, M. Ghezzo, C. Montoneri and R. Mendichi, *Waste Management*, 2011, **31**, 10.

10. P. Savarino, E. Montoneri, G. Musso and V. Boffa, *J. Surfact. Deter.*, 2010, **13**, 59.

11. P. Savarino, E. Montoneri, S. Bottigliengo, V. Boffa, T. Guizzetti, D. G. Perrone and R. Mendichi, *Ind. Eng. Chem. Res.*, 2009, **48**, 3738.

12. V. Boffa, D. G. Perrone, E. Montoneri, G. Magnacca, L. Bertinetti, L. Garlasco and R. Mendichi, *ChemSusChem*, 2010, **3**, 445.

13. R. Bordes, M. Vedrenne, Y. Coppel, S. Franceschi, E. Perez and I. Rico-Lattes, *ChemPhysChem.*, 2007, **8**, 2013.

14. M. Patel and B. K. Padhi, *J. Mater. Sci.*, 1990, **25**, 1335.

15. T. 4. Fan, S. K Chow and D. Zhang, *Prog. Mater. Sci.*, 2009, **54**, 542.

16. P. Greil, *J. Eur. Ceram. Soc.*, 2001, **21**, 105.

17. A. Zampieri, S. Kullmann, T. Selvam, J. Bauer, W. Schwieger, H. Sieber, T. Fey and P. Greil, *Micropor. Mesopor. Mater.*, 2006, **90**, 162.

18. J. Cao, C. Rambo and H. Sieber, *Ceram. Int.*, 2004, **30**, 1967.

19. J. Cao, C. Rambo and H. Sieber, *J. Porous Mater.*, 2004, **11**, 163.

20. V. P. Valtchev, M. Smaihi and A. C. Faust, *Chem. Mater.*, 2004, **16**, 1350.

21. M. Patel and B. K. Padhi, *J. Mater. Sci. Lett.*, 1993, **12**, 1234.

22. J. Cao, O. Rusina and H. Sieber, *Ceram. Int.*, 2004, **30**, 1971.

23. N. S. Venkataramanan, K. Matsui, H. Kawanami and Y. Ikushima, *Green Chem.*, 2007, **9**, 18.

24. Z. Miao, K. Ding, T. Wu, Z. Liu, B. Han, G. An, S. Miao and G. Yang, *Micropor. Mesopor. Mater.*, 2008, **111**, 104.

25. Y. Zhang, L. Hu, J. Han, Z. Jiang and Y. Zhou, *Micropor. Mesopor. Mater.*, 2010, **130**, 327.

26. W. He, J. Cui, Y. Yue, X. Zhang, X. Xia, H. Liu and S. Lui, *J. Colloid Interface Sci.*, 2011, **354**, 109.

27. X. Tian, W. He, J. Cui, X. Zhang, W. Zhou, S. Yan, X. Sun, X. Han, S. Han and Y. Yue, *J. Colloid Interface Sci.*, 2010, **343**, 344.

28. D. Yang, L. Qi and J. Ma, *Adv. Mater.*, 2002, **14**, 1543.

29. A. R. Maddocks and A. T. Harris, *Mater. Lett.*, 2009, **63**, 748.

30. A. Zampieri, G. T. P. Mabande, T. Selvam, W. Schwieger, A. Rudolph, R. Hermann, H. Sieber and P. Greil, *Mater. Sci. Eng. C*, 2006, **26**, 130.

31. M. F. Desimone, C. Helary, I. B. Rietveld, I. Bataille, G. Mosser, M. M. Giraud-Guille, J. Livage and T. Coradin, *Acta Biomater.*, 2010, **6**, 3998.

32. D. G. Perrone, V. Boffa, G. Magnacca and E. Montoneri, accepted for publication in *Mater. Sci. Applic.*

33. G. Magnacca, E. Laurenti, E. Vigna, Franzoso, L. Tomasso, E. Montoneri and V. Boffa, *Process Biochem.*, 2012, **47**, 2025.

34. R. Knechtel, *Microsyst.Technol.*, 2005, **12**, 63.
35. K. Nötzold, J. Graf and R. Müller-Fiedler, *Microelectron. Reliab.*, 2008, **48**, 1562.
36. Y. Zhang, Y. Yang, J. Zheng, W. Hua and G. Chen, *Mater. Chem. Phys.*, 2009, **114**, 319.
37. J. H. Yi, H. Y. Koo, J. H. Kim, Y. N. Ko, Y. J. Hong, Y. C. Kang and H. M. Lee, *J. Alloys Compd.*, 2011, **509**, 6325.
38. A. R. Sheldon, *Adv. Synth. Catal.*, 2007, **349**, 1289.
39. M. Chaplin, C. Bucke, *Enzyme Technol.*, 1990, Cambridge University Press, U.K.
40. L. Cao, *Curr. Op. Chem. Biol.*, 2005, **9**, 217.
41. T. Marchis, P. Avetta, A. Bianco Prevot, D. Fabbri, G. Viscardi and E. Laurenti, *J. Inorg. Biochem.*, 2011, **105**, 321.
42. A. Henriksen, O. Mirza, C. Indiani, K. Teilum, G. Smulevich, K. Welinder and M. Gajhede, *Protein Sci.*, 2001, **10**, 108.
43. H. B. Dunford, *Heme Peroxidases*, John Wiley & Sons Eds. New York, USA, 1999.
44. T. Marchis, G. Cerrato, G. Magnacca, V. Crocellà and E. Laurenti, *Biochem. Eng. J.*, 2012, **67**, 28.
45. T. T. Ngo and H. M. Lenhoff, *Anal. Biochem.*, 1980, **105**, 389.
46. P. Quagliotto, E. Montoneri, F. Tambone, F. Adani, R. Gobetto and G. Viscardi, *Environ. Sci. Technol.*, 2006, **40**, 1686.
47. S. Canonica, *Chimia*, 2007, **61**, 641.
48. W. J. Cooper, R. G. Zika, R. G. Petasne and J. M. C. Plane, *Environ. Sci. Technol.*, 1988, **22**, 1156.
49. J. V. Goldstone and B. M. Voelker, *Environ. Sci. Technol.*, 2000, **34**, 1043.
50. A. Paul, S. Hackbarth, R. D. Vogt, B. Röder, B. K. Burnison and C. E. Steinberg, *Photochem. Photobiol. Sci.*, 2004, **3**, 273.
51. P. P. Vaughan and N. V. Blough, *Environ. Sci. Technol.*, 1998, **32**, 2947.
52. T. E. Thomas-Smith and N. V. Blough, *Environ. Sci. Technol.*, 2001, **35**, 2721.
53. A. J. Elliot, *Int. J. Radiat. Applic. Instrum. Part C. Radiat. Phys. Chem.*, 1989, **34**, 753.
54. W. R. Haag and J. Hoigne, *Environ. Sci. Technol.*, 1986, **20**, 341.
55. W. R. Haag, J. Hoigne, E. Gassman and A. M. Braun, *Chemosphere*, 1984, **13**, 641.
56. M. A. J. Rodgers and P. T. Snowden, *J. Am. Chem. Soc.*, 1982, **104**, 5541.
57. L. Carlos, B. W. Pedersen, P. R. Ogilby and D. O. Martire, *Photochem. Photobiol. Sci.*, 2011, **10**, 1080.
58. R. M. Cory, J. B. Cotner and K. McNeill, *Environ. Sci. Technol.*, 2008, **43**, 718.
59. D. P. Hessler, F. H. Frimmel, E. Oliveros and A. M. Braun, *J. Photochem. Photobiol. B: Biology*, 1996, **36**, 55.
60. R. A. Larson and E. J. Weber, *Carbohyd. Polym.*, 1996, **29**, 293.
61. C. Richard, D. Vialaton, J. P. Aguer and F. Andreux, *J. Photochem. Photobiol. A: Chemistry*, 1997, **111**, 265.

62. R. M. Dalrymple, A. K. Carfagno and C. M. Sharpless, *Environ. Sci. Technol.*, 2010, **44**, 5824.
63. I. B. Afanas'ev, *Superoxide Ion: Chemistry and Biological Implications.* CRC Press: Boca Raton, FL, 1989. Vol. 1.
64. B. H. J. Bielski and A. O. Allen, *J. Phys. Chem.*, 1977, **81**, 1048.
65. O. Furman, D. F. Laine, A. Blumenfeld, A. L. Teel, K. Shimizu, I. F. Cheng and R. J. Watts, *Environ. Sci. Technol.*, 2009, **43**, 1528.
66. L. Carlos, M. Cipollone, D. B. Soria, M. Sergio Moreno, P. R. Ogilby, F. S. García Einschlag and D. O. Mártire, *Sep. Purific. Technol.*, 2012, **91**, 23.
67. W. J. Cooper, R. G. Zika, R. G. Petasne and J. M. C. Plane, *Environ. Sci. Technol.*, 1988, **22**, 1156.
68. W. R. Haag and J. Hoigne, *Chemosphere*, 1985, **14**, 1659.
69. S. E. Page, W. A. Arnold and K. McNeill, *Environ. Sci. Technol.*, 2011, **45**, 2818.
70. S. E. Page, M. Sander, W. A. Arnold and K. McNeill, *Environ. Sci. Technol.*, 2012, **46**, 1590.
71. F. L. Rosario-Ortiz, S. P. Mezyk, D. F. R. Doud and S. A. Snyder, *Environ. Sci. Technol.*, 2008, **42**, 5924.
72. P. Boule, M. Bolte, C. Richard, *Handbook of Environmental Chemistry*, 1999, Vol. 2, Part 1: Environmental Photochemistry; Boule, P. Ed.; Springer: Berlin Heidelberg, 204–205.
73. R. G. Zepp, P. F. Schlotzhauer and R. M. Sink, *Environ. Sci. Technol.*, 1985, **19**, 74.
74. S. Canonica, B. Hellrung, P. Müller and J. Wirz, *Environ. Sci. Technol.*, 2006, **40**, 6636.
75. S. Canonica, U. Jans, K. Stemmler and J. Hoigné, *Environ. Sci. Technol.*, 1995, **29**, 1822.
76. Y. Chen, C. Hu, X. Hu and J. Qu, *Environ. Sci. Technol.*, 2009, **43**, 2760.
77. J. J. Guerard, Y. P. Chin, H. Mash and C. M. Hadad, *Environ. Sci. Technol.*, 2009, **43**, 8587.
78. J. J. Werner, K. McNeill and W. A. Arnold, *Chemosphere*, 2005, **58**, 1339.
79. H. Xu, W. J. Cooper, J. Jung and W. Song, *Water Res.*, 2011, **45**, 632.
80. P. M. David Gara, G. N. Bosio, V. B. Arce, L. Poulsen, P. R. Ogilby, R. Giudici, M. C. Gonzalez and D. O. Martire, *Photochem. Photobiol.*, 2009, **85**, 686.
81. S. Malato, P. Fernández-Ibáñez, M. I. Maldonado, J. Blanco and W. Gernjak, *Catal. Today*, 2009, **147**, 1.
82. S. K. Khetan and T. J. Collins, *Chem. Rev.*, 2007, **107**, 2319.
83. D. Bahnemann, *Sol. Energy*, 2004, **77**, 445.
84. J. J. Pignatello, E. Oliveros and A. MacKay, *Crit. Rev. Environ. Sci. Technol.*, 2006, **36**, 1.
85. S. Malato, J. Blanco, D. C. Alarcón, M. I. Maldonado, P. Fernández-Ibáñez and W. Gernjak, *Catal. Today*, 2007, **122**, 137.
86. M. L. Marín, A. Arques, L. Santos-Juanes, A. M. Amat and M. A. Miranda, *Chem. Rev.*, 2012, **112**, 1710.

87. N. Klamerth, N. Miranda, S. Malato, A. Agüera, A. R. Fernández-Alba, M. I. Maldonado and J. M. Coronado, *Catal. Today*, 2009, **144**, 124.
88. A. A. Khodja, O. Trubetskaya, O. Trubetskoj, L. Cavani, C. Ciavatta, G. Guyot and C. Richard, *Chemosphere*, 2006, **62**, 1021.
89. A. Bianco Prevot, D. Fabbri, E. Pramauro, C. Baiocchi, C. Medana, E. Montoneri and V. Boffa, *J. Photochem. Photobiol. A: Chem.*, 2010, **209**, 224.
90. P. Avetta, A. Bianco Prevot, D. Fabbri, E. Montoneri and L. Tomasso, *Chem. Eng. J.*, 2012, **197**, 193.
91. A. Bianco Prevot, P. Avetta, D. Fabbri, E. Montoneri, A. Morales-Rubio and M. de la Guardia, *Anal. Bioanal. Chem.*, 2012, **404**, 657.
92. A. Bianco Prevot, D. Fabbri, E. Pramauro, C. Baiocchi and C. Medana, *J. Chromatogr. A*, 2008, **1202**, 145.
93. J. Rivera-Utrilla, M. Sánchez-Polo and C. A. Zaror, *Phys. Chem. Chem. Phys.*, 2002, **4**, 1129.
94. O. Zerbinati, M. Vincenti, S. Pittavino and M. C. Gennaro, *Chemosphere*, 1997, **35**, 2295.
95. S. Riediker, M. J. F. Suter and W. Giger, *Water Res.*, 2000, **34**, 2069.
96. D. Fabbri, A. Bianco Prevot and E. Pramauro, *J. Appl. Electrochem.*, 2005, **35**, 815.
97. E. Montoneri, *Biochemenergy*. www.biochemenergy.it (accessed February 2, 2012).
98. M. Zhan, X. Yang, Q. Xian and L. Kong, *Chemosphere*, 2006, **63**, 378.
99. M. S. Alam, B. S. M. Rao and E. Janata, *Radiat. Phys. Chem.*, 2003, **67**, 723.
100. A. Janczyk, E. Krakowska, G. Stochel and W. Macyk, *J. Am. Chem. Soc.*, 2006, **128**, 15574.
101. S. Halladja, A. Ter Halle, J. P. Aguer, A. Boulkam and C. Richard, *Environ. Sci. Technol.*, 2007, **41**, 6066.
102. J. P. Aguer, C. Richard and F. Andreux, *J. Photochem. Photobiol. A: Chemistry*, 1997, **103**, 163.
103. A. Bianco Prevot, P. Avetta, D. Fabbri, E. Laurenti, T. Marchis, D. G. Perrone, E. Montoneri and V. Boffa, *ChemSusChem*, 2011, **4**, 85.
104. I. D. Campbell and R. A. Dwek, *Biological Spectroscopy*, Benjamin/Cummings Publishing Company, USA, 1984.
105. A. Bianco Prevot, P. Avetta, D. Fabbri, D. G. Perrone, E. Montoneri and V. Boffa, *Desalin. Water Treat. (DWT) J.*, 2012, **39**, 308.
106. M. Negre, C. Mozzetti Monterumici, D. Vindrola, G. Piccone, D. G. Perrone, L. Tomasso and E. Montoneri, *Compost. Sci. Util.*, 2012, **20**, 150.
107. O. Sortino, M. Dipasquale, E. Montoneri, L. Tomasso, D. G. Perrone, D. Vindrola, M. Negre and G. Piccone, *Waste Manag.*, 2012, **32**, 1792.
108. O. Sortino, M. Dipasquale, E. Montoneri, L. Tomasso, P. Avetta and A. Bianco Prevot, *Agron. Sustain. Develop.*, 2012, in press, DOI: 10.1007/s13593-012-0117-6.

CHAPTER 5
Uses of Waste Starch

PETER S. SHUTTLEWORTH*[a] AND
NONTIPA SUPANCHAIYAMAT[b]

[a] Departamento de Física de Polímeros, Elastómeros y Aplicaciones
Energéticas Instituto de Ciencia y Tecnología de Polímeros, CSIC, Calle
Juan de la Cierva 3, 28006 Madrid, Spain; [b] Green Chemistry Centre of
Excellence, Chemistry Department, University of York, Heslington, York,
United Kingdom, YO105DD
*Email: peter@ictp.csic.es

5.1 Introduction

Starch, one of the most abundant polysaccharides and the main storage supply
for botanical resources,[1] plays an intricate role in our societies' structure as
a whole.

The United Kingdom's annual starch use equates to some 880,000 tonnes,
with only 25% of this being used in nonfood applications (paper, detergents
and plastics industry, *etc.*), with the rest adopted for both human and animal
consumption.[2,3]

Of the whole EU potato industry, the UK has an approximate 14.5% share,
from which it is estimated that there are some $10–18\,\text{kT}\,\text{y}^{-1}$ of byproducts that
consist of recoverable forms of starch (including peels, *etc.*, some of which are
supplemented in animal feed).[4] In addition, of the potato genotypes grown for
consumption, it is estimated that 8% due to size deficiencies are not used,
resulting in a considerable $530\,\text{kT}\,\text{y}^{-1}$, at a cost to industry of $£26.5\text{m}\,\text{y}^{-1}$.[4] The
wastage of starch in this industry only highlights a small fraction of the total
starch industry. Therefore, it can be seen that there is a large underutilised
resource available, *i.e.* starch slurries; that are in most cases ideal for use in the

RSC Green Chemistry No. 24
The Economic Utilisation of Food Co-Products
Edited by Abbas Kazmi and Peter Shuttleworth
© The Royal Society of Chemistry 2013
Published by the Royal Society of Chemistry, www.rsc.org

enzymatic conversion to high-value chemicals or gelatinisation procedures, *etc.* as will be discussed later.

5.2 Starch Structural Characteristics

Starch is a carbohydrate source that is used for energy storage in plants. It is stored in granular form, and used for growth, germination and in times of dormancy.[5] Chemically the composition of the starch granule varies depending on the origin and type of plant. Nevertheless, 97–99% of the granule is composed of the alpha-glucans amylose and amylopectin, with trace amounts of proteins, lipids, phosphorus, and other inorganic materials with a moisture content of about 10–20%.[6] The ratio of these polymers within the granule has a great deal of influence on physical properties such as gel formation, viscosity, solubility and strength of films formed.[7]

Amylose is defined as a linear polymer consisting of approximately 99% $(1 \rightarrow 4)$ linked α-D-glucopyranosyl units, and 1% $(1 \rightarrow 6)$ linked α-D-glucopyranosyl branched points.[8] The molecular weight is typically quoted as being in the range of 1×10^5–1×10^6 g mol^{-1}, with a degree of polymerisation by number (DP$_n$) of 324 for high amylose corn starch (amylomaize), to 4920 for potato starch, and containing between 9 and 20 branch points per molecule.[9–10] Each chain of the polymer ranges from 200 to 700 glucose residues, which is equivalent to a molecular weight of 32, 400 to 113, 400 Da.[8] These characteristics of the polymer have been determined with a number of techniques including viscosity measurements and GPC *etc.* The linear structure of amylose is shown in Figure 5.1.

Amylose can vary in proportion from 0 to 70%, depending on the starch type, but typically it has a concentration in the range 25–30% w/w.[11–12]

Amylopectin is a highly branched molecule consisting of relatively short chains of $(1 \rightarrow 4)$ - linked α-D-glucopyranosyl, containing approximately 5–6% of α-$(1 \rightarrow 6)$ glycosidic branch points as shown in Figure 5.2.[13]

Amylopectin is a much larger molecule than amylose with a molecular weight in the order of 1×10^7–1×10^9 g mol^{-1}. It is made up of three main species with degrees of polymerisation (DP) in the range of 13, 400–26, 500, 4400–8400 and 700–2100.[14] The individual chains are classified in terms of their length

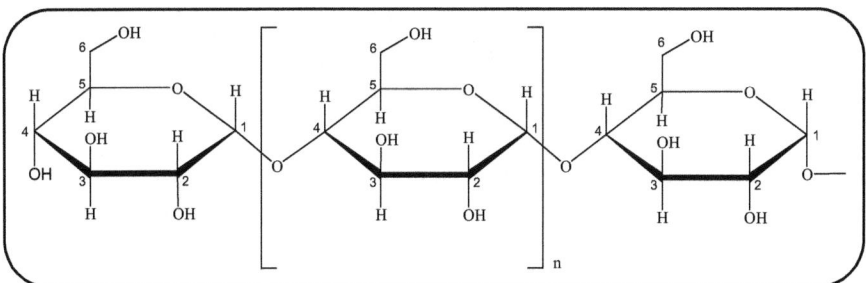

Figure 5.1 Amylose α-$(1 \rightarrow 4)$ - glucan.

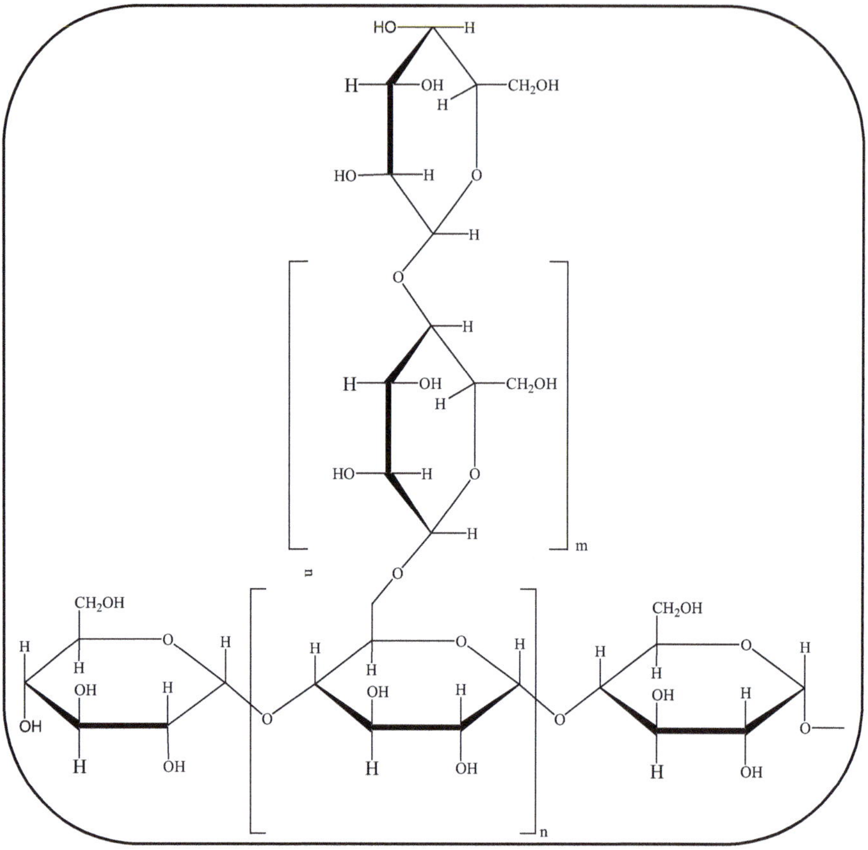

Figure 5.2   Amylopectin molecule with (1 → 6) branch points shown.

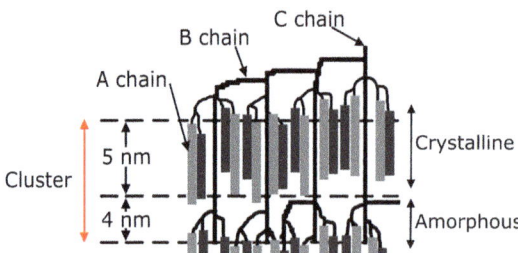

Figure 5.3   Amylopectin chains arranged in a cluster type structure.

and positioning within the starch granules and are believed to be arranged in a cluster type arrangement proposed by Robin *et al.* and French (see Figure 5.3).[9,15–17]

The exterior chains of the amylopectin molecule have been classified as A and B chains. The A chains are unbranched except for a single linkage to a B chain, whereas B chains are branched containing more than two linkages, connected to either A, B or C chains. Each amylopectin molecule contains one C chain, which has a reducing end and is said to be the backbone of the amylopectin molecule.[18]

B chains within clusters are assigned as B_1–B_4 and are designated according to their chain length. Chain lengths of the corresponding chains A, and B_1–B_4 are 12–16, 20–24, 42–48, 69–75 and 101–119, respectively, with the molar distribution of A to B -chains, being 1:1 to 1:2, depending on starch type.[6]

Native starches are semicrystalline structures. This concept has been extended to describe the levels of structural complexity that are now regarded as being important to the granule structure. The starch granule is composed of alternating regions of ordered (crystalline lamella), and less-ordered (amorphous lamella) regions. The ordered regions are composed of tightly packed, parallel glucan chains, whereas the less-ordered regions are mainly composed of branched B - chains. The size of each cluster is around 9 nm, with the crystalline region being approximately 5 nm.[19]

The crystallinity of these regions is classified according to X-ray diffraction patterns, and assigned as having A- or B-type crystallinity, or possibly an intermediate C type,[20] with A-type crystallinity being the most densely packed. Cereal starches usually have an A-type X-ray diffraction pattern, while tuber and other types of starches have B- or C-type crystallinity.

Crystallinity in starch occurs within the ordered segments of the amylopectin molecule. When the chains of these molecules are 10 or more glucose units long, they can intertwine forming a double helix arrangement. These double helices can be formed from adjacent branches within the same amylopectin cluster, or from chains of different clusters forming a three-dimensional network. The way in which the double helices are intertwined, and the length of the A-type amylopectin chains, determines its crystal structure, and hence, what crystallinity category it belongs to, *i.e.* A, B or C. Short A chains are associated with A-type crystallinity, longer A chains display B-type crystallinity, and A chains of intermediate length show C-type crystallisation.[18]

The degree of crystallinity (15–45%) in starch, however, cannot explain its relative stability, as there is simply not enough crystalline material within the granule. It is therefore postulated, from studies on potato starch, that the crystalline domains form continuous networks of left-handed helices, in which the voids of these superhelixes are assumed to be empty, and to be approximately 8 nm wide.[21,22]

The interpenetrating superhelices are assumed to form a skeleton for the granule to develop on. This model is only demonstrated for potato starch, but it is assumed that a similar model is valid for other starches, as structural designs in nature tend to be conserved.

Superhelices are believed to be "blocklet" structures, corresponding to the framework of growth rings seen in starch granules.[23,24] These blocklets are composed of crystalline and amorphous lamellae from amylopectin, and are believed to be continuous throughout the granule.[25] The crystalline growth ring consists of normal blocklets, while the amorphous growth rings consist of defective blocklets. These are also postulated to occur where there are surface pores.

Growth rings can be seen in large starch granules with a light microscope. The blocklets can also be seen using the same technique after partial hydrolyses. From the growth ring pattern it is thought that the initial growth of the granule starts at the hilum (centre of granule), which contains the reducing ends of the amylose and amylopectin, and is packed in a less-organised manner than the rest of the granule. The continuing growth of the granule occurs with amylose and amylopectin chains radiating outwards to the surface, allowing further glucose residues to be deposited, leading to the continued growth of the starch molecules. The formation of growth rings in the development of the granule suggests that the starch material is deposited in a diurnal rhythm, as the granule increases in size.

The interaction between amylose and amylopectin in the granule is an area of great debate, as there are many aspects that are not fully understood. Generally, it is thought that the majority of amylose within the granule acts as a separate entity to the rest of the molecules. This is especially the case for shorter amylose molecules near the periphery of the granule that can leach out of the granule during the gelatinisation process. However, long-chain amylose present can form amylose–lipid inclusion complexes, double-helical structures with other amylose molecules or with amylopectin chains, or even single-helical-type structures.[6] This interaction with the amylopectin clusters increases the size of the crystalline regions but, by doing so it affects the packing efficiency of the amylopectin chains.[17] Two mechanisms have been proposed to explain this occurrence. 1.) The crystallisation of amylose and amylopectin (see Figure 5.3), suggests that increasing amylose content in A-type crystallinity may result in interference of the packing of the mesogens.[26] 2.) The interaction of the amylopectin branch points in the amorphous regions of the cluster with amylose. This will strain the main amylopectin C chain disrupting the crystalline packing of the A chains.

5.3   Applications of Starch

5.3.1   Starch Adhesives

5.3.1.1   Background

The earliest known records of starch adhesives being used was for bonding papyrus strips by Egyptians nearly 6000 years ago. Starch is still currently used, but to a lesser extent than dextrins. These were supposedly serendipitously discovered in a fire in Dublin where the partially burned starch, "pyrodextrin" was discovered. This brown sticky compound, with lower molecular weight compared to that of starch, can be pasted at higher concentrations, and hence dries more quickly, gives faster tack (resistance to separation of two materials brought momentarily into contact), and consequently increased production rates when used for attaching bottle labels, *etc.*[27,28]

5.3.1.2   Use of Additives in Starch/Dextrin Adhesives

Starch and dextrin adhesives are usually combined with a number of modifiers that impart additional desired characteristics to the adhesive.

Tackifiers such as sodium tetraborate (borax), NaOH and sodium meta-borate increase the speed at which the adhesive can stick to a substrate and can also increase bond strength.[29] They also impart increased viscosity and viscosity stability, giving better fluid properties.[30,31] Excessive use of additives such as borax, however, can result in extreme cohesiveness leading to processing problems.

Plasticisers are added to improve the finished film characteristics, increase film flexibility and the ease of processing. There are three main types: chemicals that solidify with the material, humectants such as glycerol that regulate the moisture content of the film and fatty compounds that lubricate the material. They include soaps, polyglycols, fatty alcohols, *etc.*, and have the advantage of imparting permanent film flexibility, but can only be added at low levels as they reduce bond strength and cohesiveness of the adhesive.[30]

Inert fillers such as clays and bentonite reduce the cost of the adhesive substantially but can also reduce penetration of the adhesive in the substrate and, therefore selection must be carefully made taking particle size and dispersibility into consideration.

If colouration is important, certain bleaches can be added such as sodium bisulfite, hydrogen peroxide and sodium perborate. However, such additives can affect the chemical stability of the adhesive.

Preservatives are added to control microbial growth in starch-based adhesives. This is particularly important as any deterioration can dramatically affect service life. In the past, formaldehyde and chlorinated hydrocarbons have been used, but now the focus is very much into using natural substances to impart this effect.

Modification of starch alters the properties of the material. Modification can be classed as degradative as in the case for dextrinisation (as mentioned previously), or nondegradative when acetylating etc. Degradative methods such as acid hydrolysis, oxidation and thermal processes are usually carried out with the purpose of reducing viscosity of the product for higher paste concentration with good flow and dispersion properties.[32] Nondegradative methods that chemically modify the material through ester or ether linkages change the final properties by increasing stability. Modification of the hydroxyl groups decreases the tendency of the starch fragments to retrograde or age. They can also increase the hydrophobicity of the material, broadening their use while increasing the resistance to shear. However, such treatments usually reduce tack and increase the price per kilogram of the material.

5.3.1.3   General Starch Adhesives

Starch- and dextrin-based adhesives are commonly water based and used to bond various paper and cardboard products.[33] These include envelope adhesives, wallpaper paste and cardboard box sealants, etc.[34] The use of starch based adhesive for bonding wood is also an area where these adhesives have been used traditionally and where recent development has been concerned.[35–37] However, the majority of starch-based adhesives are used within the paper and textiles industry as binders and sizing agents.[30]

A summary of major markets include the paper and corrugating industries, solid fibre laminates, wound paper tubes, grocery bags, textiles, and briquettes.[38]

5.3.2 Waste Starch Adhesive Use for Carpet-Tile Recycling

Disposal of postconsumer carpet tiles and waste from manufacture such as offcuts produces large volumes of nonreadily degradable waste for landfill, such as bitumen, PVC, SBR-latex and nylon. For example, the total amount used in Japan (2003) was $1.7 \times 10^6 \, m^2$,[39] and the estimated trade in the EEA was 45 million m^2 (2010) or over 100 million kg per annum, with 65 million kg per annum ending postconsumer in landfill or being incinerated.[40] It can therefore be seen that how these tiles are disposed of is a serious environmental concern with pressures to rectify these issues coming from manufacturers, governments and from the general public.[39]

The two main alternative treatments for land-filled carpet tiles are by recovering energy from the tile by incineration or by mechanically recycling.[41] Incineration can only happen in regions with the necessary infrastructure, and with increasing environmental pressures and governmental legislation, this option will be uncertain in the future.[42] The UK originally signed up to the Kyoto Protocol and adopted a domestic goal to reduce carbon dioxide emissions. This was reinforced in 2003 with a longer-term commitment to reducing carbon dioxide emissions by 60% by 2050, with real progress to be made by 2020.[43]

Mechanical recycling is effectively changing waste material into useful material and is becoming a favourable option. Carpet tiles can be disposed of or recycled by a number of different methods: Rejuvenating the tile,[44] adding more backing layer,[45] chopping and downgrading tiles into roofing materials, *etc.*, or into more backing layer.[46] A more sophisticated recycling method for carpet tiles involves the mechanical separation of the main layers. Effectively the fabric and backing layers are pulled apart producing two product streams. This works in principle but the efficiency in terms of energy used and the quality of the two layers once separated is questionable. The fabric or the backing layers rarely stay in one piece, also, the adhesive used to bind the two layers is still present within the fabric fibres, preventing recycling. Controlling the contamination level is the most important obstacle in the increased use of recycled fibre.[47]

A novel "Switchable" adhesive recently developed utilising waste starch has shown great potential, ensuring little or no adhesive presence within the fibre matrix, allowing both the backing layer and the fibre layer to be fully recycled, and hence preventing the need to downgrade the materials.[40] The bitumen

Figure 5.4 Representation of use of a switching waste starch-based adhesive to facilitate facile separation, allowing full recyclability of components within a closed-loop system.

backing layer could be either chopped and melted or melted and sieved to recover the glass scrim. This could then be used without any composting steps as the correct amount of filler, *etc.*, would already be contained in the material. The fabric layer, after separation, could be shredded and sold to needle-felt manufacturers or further processed. This would entail the nylon being separated from the spun-bonded polyester primary fabric and reprocessed to form yarn for the carpet-tile fabric layer. This would see the complete recycling of the carpet tile without the need to downgrade any of the components, as shown in Figure 5.4. Besides, in an increasingly environmentally conscious society, a reusable system offers many advantages for end-users to recycle.[48]

5.3.3 Waste Starch-Rich Water from Potato Processing

During potato processing, which involves different activities, such as cleaning, peeling and cutting, water is continuously required. As a result, a large amount of wastewater is generated from this process. According to Russ and Meyer-Pittroff, the potato processing industry has a specific waste index (the mass of accumulated waste divided by the mass of the saleable product) of 0.3–0.5, which is relatively high and comparable to that of the meat and sugar industries. This waste contains a high amount of protein and starch.[49] Starch in this wastewater exists in the form of free suspended solids, which can be separated using a variety of methods. Common water–starch separation techniques include the use of hydrocyclones and centrifuges.[50] Moreover, membrane technologies, such as membrane bioreactors and microfiltration, has increased their popularity in wastewater treatment.[51] The use of membranes has also exhibited promising results in recovering starch from wastewater treatment in starch processing.[52] The membrane module configuration can influence the efficiency of the filtration process as Meindersma *et al.* demonstrated that tubular module configurations gave better results than those of flat and spiral configurations.[53]

Devereux *et al.* investigated different techniques of starch separation from wastewater and found that centrifugation was the most effective technique compared to sedimentation and filtration. The recovered starches, which were further chemically and physically characterised, demonstrated similar properties to their counterparts.[50] Mironesco analysed wastewater from the production of potato snacks and reported that around 0.7% of protein and 1.66% of starch had a low microbial charge. The starch was then separated through protein precipitation, followed by filtration. Also in this case, the recovered starch exhibited chemical and physical performances comparable to those of commercial starches.[54]

5.3.4 Waste Starch to Mesoporous Carbonaceous Materials – Starbon®

The polymers amylose and amylopectin can gelatinise in water when heated to form a gel, effectively opening up the starch granule and swelling the respective polymers. On cooling, the gel undergoes a process known as retrogradation that can last a few hours to weeks depending on the ratio of the polymers (faster higher amylose content), creating a mesoporous network. The mesoporous structured starch remains intact after solvent exchanging the water for a lower surface tension solvent such as ethanol.[55] Inhouse trials have shown that there are no discernible differences between the textural properties (surface area and porosity) of virgin starch sources compared to the same genotype waste starch sources. In the cases where there was some depolymerisation, generally lower gelatinisation temperatures and faster retrogradation times were required. In addition, a large proportion of industrial waste starch is in the form of aqueous slurries that is an added benefit to the process, due to the initial step requiring mixing starch with water to a concentration between 5 to 10% w/v.

Utilising the already created mesoporous starch template and then doping with tosylic acid and heating under vacuum produces a line of materials trademarked as Starbon®. This line of materials range in functionality from near-starch like with rich hydroxyl/carboxylic acid groups, through to conjugated forms to aromatic functionalities through to graphitic textural varieties. These materials with tuneable surface chemistries and good porosity and hence, good diffusion properties have shown real promise in the areas of catalysis, adsorption and chromatography.[55–60]

5.3.5 Starch in the Paper Industry

Nonfood applications of starch present around 40% of the total starch consumption in the European Union with the paper industry dominating this sector (2.2 million tonnes, in 2008).[61] Paper is mainly composed of cellulose fibres combined with other components including pigments, binders and sizing agents.[62] Starch presents the third most utilised constituent in papermaking by weight after cellulose fibres and mineral pigments.[63]

Potato (43%), maize (38%) and wheat (18%) starches are the most commonly used varieties for paper production. Starch is utilised at different stages of this process, including surface application, wet-end addition and coating.[64] During furnish preparation, starch acts as a flocculent and retention aid and it is also used as an adhesive during surface sizing in order to bond loose segments of cellulose fibres, resulting in improved strength and stiffness of the product.[63] Starch is also commonly used in combination with latex for coating, in order to smooth out the paper surface during the drying process.[65]

Most starches utilised in papermaking are modified by controlled hydrolysis, oxidation or derivatisation. These chemical and physical modifications are required in order to obtain specific properties of viscosity, charge and bonding to fibres and pigments.[62] Controlled hydrolysis allows adjustment of viscosity of starch paste as the cleavage of the glycosidic bonds leads to shorter polymer chains in starch granules, resulting in a decrease in viscosity.[66] Cationic starch are typically modified with tertiary or quaternary ammonium compounds.[67] The modification improves the affinity of starch with cellulose fibres, which naturally carry a slight anionic charge.[68,69]

5.3.6 Starch in the Textile Industry

A large proportion of starch is used in the textile industry for sizing, finishing and printing applications,[70] warp sizing being its principal use. Modified starches are typically used, often in hot aqueous solution to be able to impregnate the assembled yarns, resulting in increased yarn strength and resistance to abrasion during weaving.[65] After weaving, all the sizing agents including starch, have to be completely removed (preparatory process) before dyeing and printing can be pursued.[71] Enzyme treatment is widely used for this step as the cellulosic textile is not affected, while the starch is converted to water-soluble compounds, which can eventually be washed off.[72] The use of ultrasound combined with the enzymatic treatment has also been reported, shortening the processing time, decreasing the amount of expensive enzyme needed, and reducing fibre damage, whilst improving the uniformity of the treatment.[73]

Starches (also modified forms) are used in the textile printing process as a thickener of the printing paste, which contains a high concentration of dye. The thickening agents are added in the dyeing paste in order to avoid capillary flow between the threads of the fabric.[74] As such, the pattern design can be printed on the fabric with good sharpness and colour. Printing thickeners are mainly produced from maize starch.[65]

To overcome some of the drawbacks of native starch, including high viscosities, large molecular size, film rigidity and susceptibility to degradation by micro-organisms, modified forms are commonly used.[75] For example, carboxymethylated starch has been intensively used in the textile industry as both a sizing and thickening agent in printing paste.[76,77]

5.3.7   Waste Starch as a Source of Chemicals

The chemical industry has relied on fossil resources for decades as a source of raw materials. However, the concept of sustainable development has recently attracted a great deal of interest in the utilisation of biomass as a novel source of chemicals. The concept biorefinery is based on the idea of replacing carbon molecules from petroleum-based chemicals by plant-extracted carbon molecules.[78] Waste starch presents a great potential as a source of bioderived chemicals.

Initially, starch is converted into simple sugars *via* a depolymerisation process, which typically involves the use of enzymes. Both α-1,4 and α-1,6-debranching hydrolases, such as endo-amylases, exo-amylases, debranching enzymes and transferases, are required in efficient starch hydrolysis.[79] Once starch is broken down, the glucose monomers can be utilised as substrates for chemical reactions leading to various chemical products. Fermentation using micro-organisms leads to different molecules, such as lactic acid, PHB and succinic acid. Direct conversion of starch to lactic acid was reported using *Lactobacillus amylovorus* and *Lactobacillus* plantarum A6.[80,81] Jin *et al.* successfully produced L(+)-lactic acid in a simultaneous saccharification and fermentation process from starch waste effluents using *Rhizopus arrhizus*, strain DAR 36017.[82] Furthermore, the same yeast was used by Zhang *et al.* to produce lactic acid from waste potato starch.[83] Lactic acid is used in various industries including food, pharmaceutical and chemical production. This molecule can also be polymerised to yield polylactic acid (PLA), which has great potential as bioderived plastic.[84] Waste potato starch can be used as a viable alternative carbon source for PHB production, using *Ralstonia eutropha* NCIMB 11599. Moreover, the production performance with the waste starch was found to be practically identical to that with glucose.[85]

Another very important glucose derivative is succinic acid, which presents one of the top 12 most useful bioderived building blocks in chemical synthesis according to the US Department of Energy (Table 5.1).[86] Succinic acid shares the basic chemistry of petrochemically derived maleic acid/anhydride. It is also considered as a versatile intermediate for the production of some of the most widely used fossil-derived chemicals (Figure 5.5). Production of succinic acid from waste starch derived from the wheat milling process using *Actinobacillus succinogenes* was reported by Dorado *et al.* This represents a great prospective for the use of waste starch as an alternative source for chemical production.[87]

5.3.8   Starch Plastics

5.3.8.1   Starch in Composites

The use of starch in composites is relatively recent compared to its uses in other applications such as adhesives, paper industry and textiles. Waste starch presents an attractive feedstock for composite materials due to its low cost, availability and biodegradability.

Table 5.1 Top 12 biomass-derived platform molecules (US Department of Energy).[86]

Four carbons 1,4-Diacids (succinic, fumaric and malic acids) R = -CH₂CH₂-, -CH=CH-, -CH₂CH(OH)-	
2,5-Furan dicarboxylic acid	
3-Hydroxypropionic acid	
Aspartic acid	
Glucaric acid	
Glutamic acid	
Itaconic acid	
Levulinic acid	
3-Hydroxybutyrolactone	
Glycerol	

Table 5.1   (*Continued*)

Sorbitol (alcohol sugar of glucose)

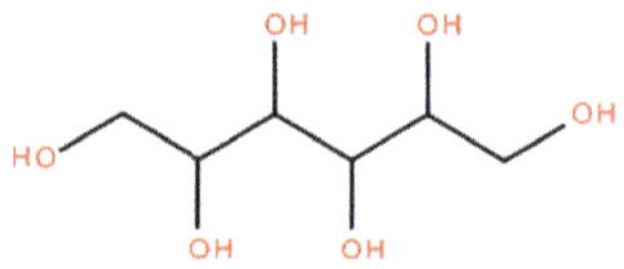

Xylitol/arabinitol (sugar alcohols)

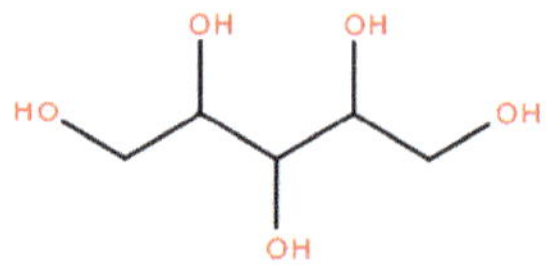

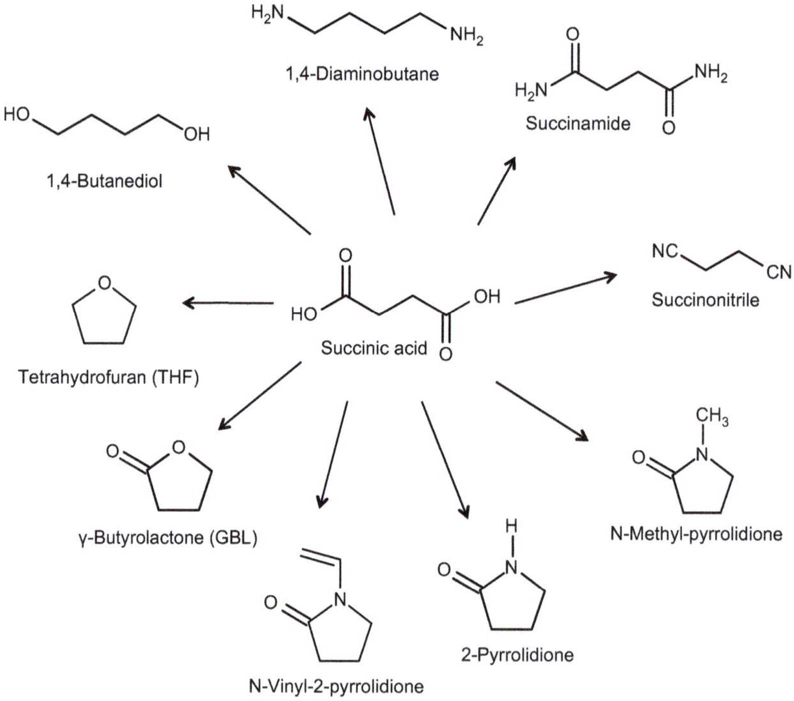

Figure 5.5   Succinic acid and its derivatives (Adapted from U.S. Department of Energy Top Value Added Chemicals from Biomass Report).[86]

5.3.8.2   *Starch Blends with Synthetic Nondegradable Polymers*

The principal objectives of blending starch with synthetic polymers are increasing the biodegradability of the materials and reducing the production cost. Partially biodegradable polymers were obtained by incorporating starch filler into polyethylene matrix.[88,89] Fillers can enhance different properties of the polymer including strength, stiffness, permeability and cost. However, poor adhesion between fillers and polymer matrix can lead to undesirable consequences, such as decreased tensile strength and elongation.[90] Starch and polyethylene are incompatible, therefore the filler or the polymer has to be modified to improve the adhesion between the two components.[91] Maleic anhydride or acrylic acid have been used as compatibilisers and grafted to polyethylene chains in order to enhance its compatibility with starch.[92] Modification of starch with n-octenyl succinic anhydride was found to increase the compatibility of starch and polyethylene, resulting in improved tensile strength of the composite.[93]

However, blending starch with nondegradable polymers only yields partially biodegradable products. In order to prepare completely biodegradable starch-based composites starch is typically blended with aliphatic polyesters, polyvinyl alcohol (PVA) or other biopolymers.

5.3.8.3 Starch Blends with Synthetic Degradable Polymers

Synthetic degradable polyesters that have been blended with starch include poly(hydroxyl alkanoates) (PHAs), poly(lactic acid) (PLA) and polycaprolactone (PCL). PHAs are naturally occurring polyesters produced by bacteria.[94] One of the most studied PHAs is poly(3-hydroxybutyrate) (PHB).[95] The primary aim of the inclusion of starch into PHB resins was to improve the properties of the composites and lower the production cost. The resin produced from PHB/starch ratio 70/30 (wt./wt.) present a cost-effective material with enhanced mechanical properties compared to those of virgin PHB.[96]

Poly(lactic acid) is a biodegradable thermoplastic polymer of lactic acid that can be obtained through waste starch/ carbohydrate fermentation (as discussed previously in Section 5.2.7).[84] Starch and PLA are immiscible due to the poor interfacial attraction between starch granules and PLA matrix. Compatibilisers such as maleic anhydride or methylenediphenyldiisocyanate (MDI) are therefore needed in the composite to maintain the mechanical properties of the product.[97,98]

Starch has also been blended with petroleum-derived biodegradable polymers such as PCL and PVA. The association of PCL with thermoplastic starch (TPS) has been studied and the dimensional stability and hydrophobicity of the blends was improved significantly compared to those of native TPS, despite the phase separation of the blend.[99] The use of high amylose starch as a filler in PCL-based materials gives better mechanical properties than other starches owing to its small granular size, allowing good dispersion in PCL matrix.[100]

PVA and starch are compatible and the addition of PVA into the starch matrix has demonstrated enhanced tensile strength and elongation.[101] However, the fact that both starch and PVA are hydrophilic limits their applications. Methods that have been attempted to enhance the water resistance of the composites including modification of starch and addition of crosslinkers.[102,103]

5.2.8.4   Starch Blends with other Biopolymers

Biopolymers such as chitosan, cellulose and pectin, have been blended with starch. Chitosan has a good film-forming property and is also compatible with starch.[104] The inclusion of starch in chitosan films improve tensile strength and elongation with regular starch showing better performance than waxy starch.[105] Microcrystalline cellulose or methylcellulose has been extruded with corn starch to produce edible films. The high ratio of cellulose enhances the tensile strength and elongation of the materials and reduces the water vapour transmission rate and gas permeability.[106]

Starch has been blended with pectin to yield edible biodegradable and water-sensitive composites. As such these materials can be used in applications such as water-soluble pouches for detergents and insecticides, edible bags for soup or noodles ingredients and drug-delivery systems.[107,108]

5.3.9   Ethanol Production from Waste Starch

The world production of ethanol has rapidly increased in the last few years, reflecting its high demand, in particular as a biofuel. The world ethanol production rose from 49.44 million litres in 2007 to 87.10 million litres in 2010, representing an increase of approximately 76%.[109] The United States of American and Brazil are the largest ethanol exporters, however the primary source of the production is maize in the USA, while the majority of ethanol in Brazil is produced from sugarcane.[110]

Ethanol is typically produced from the fermentation of sucrose in sugarcane or glucose derived from starch-based crops.[111] A yeast, *Saccharomyces cerevisiae,* is commonly used in the fermentation process to obtain bioethanol. Numerous studies have been attempted in order to increase the yield of this process, especially a metabolic engineering approach.[79] The first generation of biofuel production provokes some concerns regarding the competition with food for feedstocks and fertile lands, but use of waste starch sources should ease these concerns.[112] In this account the use of waste starch could be associated with second-generation biofuels that utilise nonfood crops as feedstocks for biofuel production.

Waste starch corn kernels from the corn seed breeding industry, which contains 63% starch, was used as a source for ethanol production. The waste stream was initially hydrolysed, followed by a fermentation process using *Saccharomyces cerevisiae* to yield upto 22.82% ethanol production.[113] Davis *et al.* reported the production of ethanol from residual starch from flour wet milling and found that *Zymomonas mobilis* ZM4 was more efficient than *Saccharomyces cerevisiae*, which is industrially used for ethanol production. This was evidenced by the higher maximum specific rates of glucose uptake and ethanol yields close to theoretical in the case of *Zymomonas mobilis* ZM4.[111]

5.4 Summary

Starch recovered from different types of waste streams represents an interesting and alternative source of starch due to its beneficial environmental aspects. It can also be cost effective if the right technology is adopted during the recovering process. It has been shown that in many cases waste starch is structurally intact and similar in properties to virgin sources, and therefore, can potentially be used for various nonfood applications with little or no concerns. These not only include traditional starch applications, such as adhesive, paper and textile uses, but also in the novel and exciting areas of catalysts or as a source of numerous chemicals currently used in industry. The conversion of waste materials into these new value-added products helps shape the path towards a more sustainable future.

Acknowledgements

The authors would like to thank past and present members of the Green Chemistry Centre of Excellence for their input and useful discussions. PS gratefully acknowledges the Ministerio de Ciencia e Innovación for the concession of a Juan de la Cierva (JCI-2011-10836) contract and the ICTP, CSIC. NS is a Thai government scholar and gratefully acknowledges the financial support of the royal Thai government.

References

1. L. Avérous, in *Polysaccharide Building Blocks*, John Wiley & Sons, Inc., Hoboken, NJ, USA, 2012, pp. 307–329.
2. G. Entwistle, S. Bachelor, E. Booth and K. Walker, *Ind. Crop. Prod.*, 1998, **7**, 175–186.
3. L. I. Ltd, *Supply Chain Impacts of Further Regulation of Products Consisting of, Containing, or Derived from, Genetically Modified Organisms*, Department for Environment, Food and Rural Affairs, 2003.
4. R. M. d. Bragança and P. Fowler, *Industrial Markets For Starch*, The BioComposites Centre, University of Wales, Bangor, 2004.
5. J. Swinkels, in *Starch Conversion Technology*, eds. V. Beynum and J. Roels, Marcel Dekker Inc., New York, 1985, pp. 15–46.
6. R. F. Tester, J. Karkalas and X. Qi, *J. Cereal. Sci.*, 2004, **39**, 151–165.
7. D. Thomas and W. Atwell, *Starches*, Eagan Press, Minnesota, 1999.
8. R. Tester and J. Karkalas, in *Biopolymers, Polysaccharides II: Polysaccharides from Eukaryotes, Volume 6*, eds. E. Vandamme, S. De Baets and A. Steinbüchel, Wiley-VCH, Weinhiem, 2002, pp. 388–390.
9. A. Buléon, P. Colonna, V. Planchot and S. Ball, *Int. J. Biol. Macromol.*, 1998, **23**, 85–112.
10. Y. Yashushi, T. Takenouchi and Y. Takeda, *Carbohyd. Polym.*, 2002, **47**, 159–167.

11. P. Orford, S. Ring, V. Carroll, M. Miles and V. Moris, *J. Sci. Food. Agr.*, 1987, **39**, 169–177.

12. S. Jobling, *Curr. Opin. Plant Biol.*, 2004, **7**, 210–218.

13. A. Blennow, M. Hansen, A. Schulz, K. Jørgensen, A. M. Donald and J. Sanderson, *J. Struct. Biol.*, 2003, **143**, 229–241.

14. Y. Takeda, S. Shibahara and I. Hanashiro, *Carbohyd. Res.*, 2003, **338**, 471–475.

15. J. Robin, C. Mercier, R. Charbonnière and A. Guilbot, *Cereal Chem.*, 1974, **51**, 389–406.

16. D. French, *J. Jpn. Soc. Starch Sci.*, 1972, **19**, 8–25.

17. P. J. Jenkins and A. M. Donald, *Int. J. Biol. Macromol.*, 1995, **17**, 315–321.

18. T. Wang, T. Bogracheva and C. Hedley, *J. Exp. Bol.*, 1998, **49**, 481–502.

19. C. Oates, *Trends Food Sci. Tech.*, 1997, **8**, 375–382.

20. R. Vereylen, B. Goderis, H. Reynaers and J. Delcour, *Biomacromolecules*, 2004, **5**, 1775–1786.

21. G. T. Oostergetel and E. F. J. van Bruggen, *Carbohyd. Polym.*, 1993, **21**, 7–12.

22. S. Ball, H. Guan, M. James, A. Myers, P. Keeling, G. Mouille, A. Buléon, P. Colonna and J. Preiss, *Cell*, 1996, **86**, 349–352.

23. D. J. Gallant, B. Bouchet and P. M. Baldwin, *Carbohyd. Polym.*, 1997, **32**, 177–191.

24. E. Bertoft, *Carbohyd. Polym.*, 2004, **57**, 211–224.

25. H. Tang, T. Mitsunaga and Y. Kawamura, *Carbohyd. Polym.*, 2006, **63**, 555–560.

26. J. Sanderson, R. Daniels, A. Donald, A. Blennow and S. Engelsen, *Polymers*, 2006, **64**, 433–443.

27. H. Kennedy and A. Fischer, in *Starch: Chemistry and Technology, 2nd revised edition*, eds. R. Whistler, J. BeMiller and E. Paschall, Academic Press Inc., London, 1984, pp. 593–610.

28. D. Aubrey, in *Handbook of Adhesion, 2nd edition*, ed. D. Packman, John Wiley & Sons, Ltd., Chichester, 2005, p. 527.

29. F. Emengo, S. Chukwu and J. Mozie, *Int. J. Adhes. Adhes.*, 2002, 93–100.

30. M. Baumann and A. Conner, in *Handbook of Adhesive Technology*, eds. A. Pizzi and K. Mittal, Marcel Dekker Inc., New York, 1994, pp. 299–313.

31. P. Thansandote, M. Raemy, A. Rudolph and M. Lautens, *Org. Lett.*, 2007, **9**, 5255–5258.

32. I. Skeist, *Handbook of Adhesives, 2nd edition*, Van Nostrand Reinhold Company, New York, 1962.

33. T. Minoru and I. Hideyuki, JP Patent 2006083336, 2006.

34. N. Yoshihiko and M. Toshihiro, JP Patent 2006117869, 2005.

35. S. Imam, S. Gordon, L. Mao and L. Chen, *Polym. Degrad. Stabil.*, 2001, **73**.

36. Z. Wang, Z. Li, Z. Gu, Y. Hong and L. Cheng, *Carbohydr. Polym.*, 2012, **88**, 699–706.

37. Z. Wang, Z. Gu, Y. Hong, L. Cheng and Z. Li, *Carbohydr. Polym.*, 2011, **86**, 72–76.
38. H. Kennedy, in *Adhesives from Renewable Resources, ACS Symposium Series 385*, eds. R. Hemingway, A. Conner and S. Branham, American Chemical Society, Washington, 1989, pp. 326–336.
39. S. Kobayashi, T. Nord and A. Suzuki, US Patent 2003/0037508, 2003.
40. P. S. Shuttleworth, J. H. Clark, R. Mantle and N. Stansfield, *Green Chem.*, 2010, **12**, 798–803.
41. A. Garforth, S. Ali, J. Hernández-Martínez and A. Akah, *Curr. Opin. Solid. State Mater.*, 2004, **8**, 419–425.
42. S. McNeil, M. Sunderland and L. Zaitseva, *Resour. Conserv. Recy.*, 2007, **51**, 220–224.
43. DEFRA, *The United Kingdom's Report on Demonstrable Progress under the Kyoto Protocol*, 2006.
44. R. Brown and K. Higgins, US Patent 2003/0203116 A1, 2003.
45. R. Brown and K. Higgins, US Patent 6,989,037 B2, 2006.
46. M. Bell, US Patent 2005/0042413 A1, 2005.
47. J. Guo, S. Severtson and L. Gwin, *Ind. Eng. Chem. Res.*, 2007, **46**, 2753–2759.
48. J. Malik and S. Clarson, *Int. J. Adhes. Adhes.*, 2002, **22**, 283–289.
49. W. Russ and R. Meyer-Pittroff, *Crit. Rev. Food. Sci.*, 2004, **44**, 57–62.
50. S. Devereux, P. Shuttleworth, D. Macquarrie and F. Paradisi, *J. Polym. Environ.*, 2011, **19**, 971–979.
51. A. Bennett, *Filtr. Separat.*, 2005, **42**, 28–30.
52. K. D. Rausch, *Starch/Stärke*, 2002, **54**.
53. G. W. Meindersma, *Starch - Stärke*, 1980, **32**, 329–334.
54. M. Mironescu, *J. Agro. Process. Tech.*, 2011, **17**, 134–138.
55. P. S. Shuttleworth, A. Matharu and J. H. Clark, in *Polysaccharide Building Blocks*, John Wiley & Sons, Inc., 2012, pp. 271–285.
56. V. Budarin, J. H. Clark, J. J. E. Hardy, R. Luque, K. Milkowski, S. J. Tavener and A. J. Wilson, *Angew. Chem.-Int. Edit.*, 2006, **45**, 3782–3786.
57. R. J. White, C. Antonio, V. L. Budarin, E. Bergstrom, J. Thomas-Oates and J. H. Clark, *Adv. Funct. Mater.*, 2010, **20**, 1834–1841.
58. R. J. White, R. Luque, V. L. Budarin, J. H. Clark and D. J. Macquarrie, *Chem. Soc. Rev.*, 2009, **38**, 481–494.
59. R. J. White, V. Budarin, R. Luque, J. H. Clark and D. J. Macquarrie, *Chem. Soc. Rev.*, 2009, **38**, 3401–3418.
60. H. L. Parker, A. J. Hunt, V. L. Budarin, P. S. Shuttleworth, K. L. Miller and J. H. Clark, *RSC Adv.*, 2012, **2**, 8992–8997.
61. Agrosynergie, *Evaluation of Common Agricultural Policy Measures applied to the Starch Sector*, European Commission, 2010.
62. H. W. Maurer, in *Starch: Chemistry and Technology, 3rd edition*, eds. J. N. Bemiller and R. L. Whistler, Academic Press, Burlington, 2009, pp. 658–713.
63. H. W. Maurer and R. L. Kearney, *Starch - Stärke*, 1998, **50**, 396–402.

64. N.-O. Bergh, in *Surface Application of Paper Chemicals*, eds. J. Brander and I. Thorn, Blackie Academic & Professional, London, 1997, pp. 69–88.

65. A. Kraak, *Ind. Crop. Prod.*, 1992, **1**, 107–112.

66. H.-J. Chung, H.-Y. Jeong and S.-T. Lim, *Carbohyd. Polym.*, 2003, **54**, 449–455.

67. S. Pal, D. Mal and R. P. Singh, *Carbohyd. Polym.*, 2005, **59**, 417–423.

68. R. P. Ellis, M. P. Cochrane, M. F. B. Dale, C. M. Duffus, A. Lynn, I. M. Morrison, R. D. M. Prentice, J. S. Swanston and S. A. Tiller, *J. Sci. Food. Agr.*, 1998, **77**, 289–311.

69. E. Sjöström, *Nordic Pulp Pap. Res. J.*, 1989, **4**, 90.

70. W. F. Breuninger, K. Piyachomkwan and K. Sriroth, in *Starch: Chemistry and Technology, 3rd edition*, eds. J. N. Bemiller and R. L. Whistler, Academic Press, Burlington, 2009, pp. 541–568.

71. A. Kumar and R. Choudhury, *Textile Preparation and Dyeing*, Science Publishers, Enfield, 2006.

72. A. S. Aly, A. B. Moustafa and A. Hebeish, *J. Clean. Prod.*, 2004, **12**, 697–705.

73. V. G. Yachmenev, N. R. Bertoniere and E. J. Blanchard, *J. Chem. Technol. Biot.*, 2002, **77**, 559–567.

74. H. Mollet and A. Grubenmann, *Formulation Technology - Emulsions, Suspensions, Solid Forms*, Wiley-VCH, Weinheim, 2001.

75. K. M. Mostafa and A. A. El-Sanabary, *Polym. Degrad. Stabil.*, 1997, **55**, 181–184.

76. M. A. El-Sheikh, *Carbohyd. Polym.*, 2010, **79**, 875–881.

77. A. A. Ragheb, H. S. El-Sayiad and A. Hebeish, *Starch - Stärke*, 1997, **49**, 238–245.

78. S. P. Octave and D. Thomas, *Biochimie*, 2009, **91**, 659–664.

79. W. van Zyl, M. Bloom and M. Viktor, *Appl. Microbiol. Biot.*, 2012, **95**, 1377–1388.

80. D. X. Zhang and M. Cheryan, *Biotechnol. Lett.*, 1991, **13**, 733–738.

81. E. Giraud, A. Champailler and M. Raimbault, *Appl. Environ. Microbiol.*, 1994, **60**, 4319–4323.

82. B. Jin, L. P. Huang and P. Lant, *Biotechnol. Lett.*, 2003, **25**, 1983–1987.

83. Z. Zhang, B. Jin and J. Kelly, *World J. Microb. Biot.*, 2007, **23**, 229–236.

84. R. Datta, S.-P. Tsai, P. Bonsignore, S.-H. Moon and J. R. Frank, *FEMS Microbiol. Rev.*, 1995, **16**, 221–231.

85. R. Haas, B. Jin and F. Tobias, *Biosci. Biotech. Bioch.*, 2008, **72**, 253–256.

86. T. Werpy and G. Peterson, *Top Value Added Chemicals Form Biomass*, U.S. Department of Energy, 2004.

87. M. P. Dorado, S. K. C. Lin, A. Koutinas, C. Du, R. Wang and C. Webb, *J. Biotechnol.*, 2009, **143**, 51–59.

88. L. Griffin Gerald J, in *Fillers and Reinforcements for Plastics*, eds. Rudolph D. Deanin and N. R. Schott, American Chemical Society, 1974, vol. 134, pp. 159–170.

89. N. St-Pierre, B. D. Favis, B. A. Ramsay, J. A. Ramsay and H. Verhoogt, *Polymer*, 1997, **38**, 647–655.
90. J. L. Willet, in *Starch: Chemistry and Technology, 3rd edition*, eds. J. N. Bemiller and R. L. Whistler, Academic Press, Burlington, 2009, pp. 715–745.
91. R. Chandra and R. Rustgi, *Polym. Degrad. Stabil.*, 1997, **56**, 185–202.
92. C.-Y. Huang, M.-L. Roan, M.-C. Kuo and W.-L. Lu, *Polym. Degrad. Stabil.*, 2005, **90**, 95–105.
93. R. L. Evangelista, W. Sung, J. L. Jane, R. J. Gelina and Z. L. Nikolov, *Ind. Eng. Chem. Res.*, 1991, **30**, 1841–1846.
94. C. Nawrath, Y. Poirier and C. Somerville, *Mol. Breeding*, 1995, **1**, 105–122.
95. C. W. Pouton and S. Akhtar, *Adv. Drug Deliver. Rev.*, 1996, **18**, 133–162.
96. S. Godbole, S. Gote, M. Latkar and T. Chakrabarti, *Biores. Technol.*, 2003, **86**, 33–37.
97. H. Wang, X. Sun and P. Seib, *J. Appl. Polym. Sci.*, 2001, **82**, 1761–1767.
98. J.-F. Zhang and X. Sun, *Biomacromolecules*, 2004, **5**, 1446–1451.
99. L. Averous, L. Moro, P. Dole and C. Fringant, *Polymer*, 2000, **41**, 4157–4167.
100. M. F. Koenig and S. J. Huang, *Polymer*, 1995, **36**, 1877–1882.
101. L. Mao, S. Imam, S. Gordon, P. Cinelli and E. Chiellini, *J. Polym. Environ.*, 2000, **8**, 205–211.
102. Z. Guohua, L. Ya, F. Cuilan, Z. Min, Z. Caiqiong and C. Zongdao, *Polym. Degrad. Stabil.*, 2006, **91**, 703–711.
103. L. Chen, S. Imam, S. Gordon and R. Greene, *J. Polym. Environ.*, 1997, **5**, 111–117.
104. T. Bourtoom and M. S. Chinnan, *LWT - Food Sci. Technol.*, 2008, **41**, 1633–1641.
105. Y. X. Xu, K. M. Kim, M. A. Hanna and D. Nag, *Ind. Crop. Prod.*, 2005, **21**, 185–192.
106. E. Psomiadou, I. Arvanitoyannis and N. Yamamoto, *Carbohyd. Polym.*, 1996, **31**, 193–204.
107. M. L. Fishman, D. R. Coffin, R. P. Konstance and C. I. Onwulata, *Carbohyd. Polym.*, 2000, **41**, 317–325.
108. K. Desai, *AAPS PharmSciTech*, 2005, **6**, E202–E208.
109. Alternative Fuels Data Center - United States Department of Energy, 2012.
110. FAO Trade and Markets Division, *Food Outlook - Global Market Analysis*, Food and Agriculture Organization of the United Nations, 2011.
111. L. Davis, P. Rogers, J. Pearce and P. Peiris, *Biomass. Bioenerg.*, 2006, **30**, 809–814.
112. F. Cherubini, *Energ. Convers. Manage.*, 2010, **51**, 1412–1421.
113. N. K. Sari, K. Y. Dharmawan and A. Gitawati, in *Bali International Seminar on Science and Technology*, Bali, Indonesia, 2011.

Food Waste and Catering Waste; Focus on Valorisation of Used Cooking Oil and Recovered Triglycerides

LORENZO HERRERO DÁVILA

Brocklesby Ltd, UK
Email: lorenzo@brocklesby.org

6.1 Valorisation of Reclaimed Oils, Fats and Triglyceride-Rich Food Coproducts

6.1.1 Context and Definitions

The issue of food waste/food supply chain waste (FW or FWSC) valorisation is becoming increasingly relevant all around the globe. The need to feed an ever-increasing human population worldwide, the unsustainability of wasting valuable natural resources in food production plus the potential for food waste as a raw material for the manufacture of renewable fuels, chemicals and materials are key drivers in political agendas worldwide.[1a,1b]

Within the industry an important source of coproducts is the world's supply chain of oils and fats. The quantification of the volumes for these coproducts is difficult to make. Taking into account global oilseed and oil yields plus oil seed meals and oil/fat production worldwide, an estimation can be made.

RSC Green Chemistry No. 24
The Economic Utilisation of Food Co-Products
Edited by Abbas Kazmi and Peter Shuttleworth
© The Royal Society of Chemistry 2013
Published by the Royal Society of Chemistry, www.rsc.org

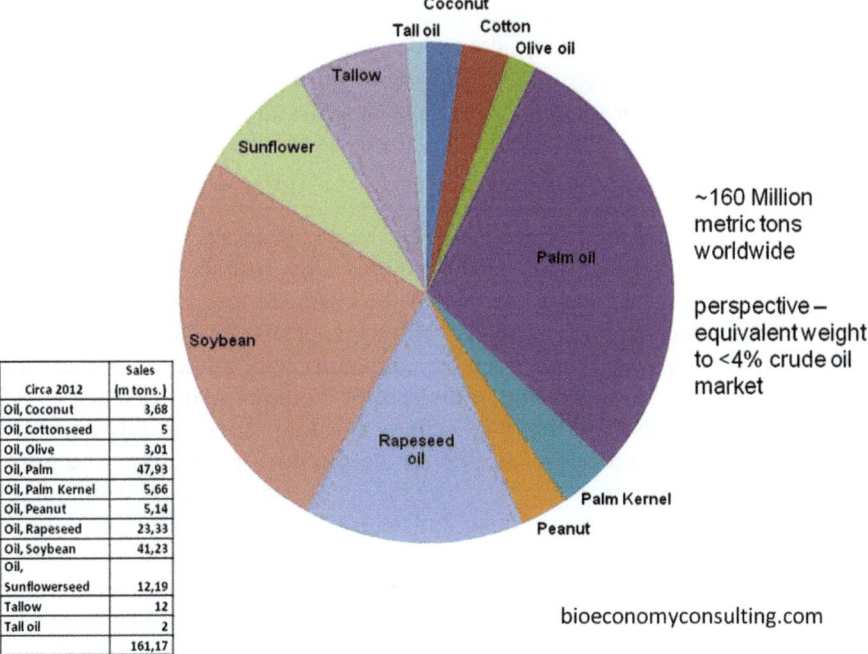

The table within the figure:

Circa 2012	Sales (m tons.)
Oil, Coconut	3,68
Oil, Cottonseed	5
Oil, Olive	3,01
Oil, Palm	47,93
Oil, Palm Kernel	5,66
Oil, Peanut	5,14
Oil, Rapeseed	23,33
Oil, Soybean	41,23
Oil, Sunflowerseed	12,19
Tallow	12
Tall oil	2
	161,17

Figure 6.1 Volumes of different oils and fats used worldwide.
(Courtesy of Mr. Aliester Ried from Bioeconomy Consulting UK, with permission).

Above 310 m tonnes of oilseeds are produced worldwide according to FAO[2] (this does not include olive fruits, at about 20 m tonnes), where world production of oils and fats is ca.160–180 million tonnes per year (Figure 6.1). These include oil pressed from oil seed, tall oils from paper manufacturing and animal fats, such as lard and tallow.

That leaves 140 m tonnes of coproducts where some 118 millions tonnes are seedmeal that tend to be used as animal feed (above 90 wt.%). Fats and oil are mainly employed within the human food supply chain (80% of the total),[3] with the remaining 20% dedicated to animal feed (5–6%), production of oleo-chemicals (surfactants, coatings and lubricants)[4] and biodiesel[5] (16.2 million metric tonnes per year being produced worldwide). Additional triglyceride-rich coproducts could be generated as a consequence of the utilisation of fats and oils in various applications.

A serious attempt to classify the coproducts generated in the oils and fats supply chain (including pre-consumer and post-consumer materials) have been made by various researchers,[6] this is important in order to assess potential for valorisation. The classification established by Bodioli primarily focuses on the potential for the use of oils and fats supply chain coproducts as an animal feed supplement by considering the following:

– identification and classification of potentially useful materials;

- knowledge of production technology;
- evaluation of chemical/physical properties;
- evaluation of the potential presence of harmful compounds;
- identification of best practices to avoid sanitary problems;
- eventual withdrawal of potentially dangerous categories to be banned for cattle feed.

In the classification the following supply chain coproducts were included:

AOCHE (Acid oils from chemical refining): a by-product of oils and fats refining process, carried out according to the chemical technology. The fatty acids are separated from the acidic oil by means of alkali addition, and then restored from the separated soaps by mineral acid addition. Obtaining AOCHEs will also yield glycerine in the case of biodiesel soap.

AOPHY (Acid oils from physical refining): a by-product of oils and fats refining process, carried out according to the physical technology. The fatty acids are separated from the acidic oil by means of vacuum/steam distillation.

LECI (Lecithins): a mixture of polar and neutral lipids are recovered from seed oils (corn, rape, sunflower, soy) immediately after extraction by water or steam.

RECY (Recycled cooking oils): Used cooking oil has traditionally been valorised into animal feed products. This is still the case only in fully traceable types of oils such as those that have not been in contact with meat products, generally known as factory vegetable oil (FVO), by contrast to catering oils which are generally designated as UCO. This differentiation is a consequence of animal by-product regulations that drove radical changes in legislation for the reutilisation of catering UCO.[7] Consequently, UCO has to be valorised to non-feed applications, distinct from the human food supply chain. UCO and FVO are sourced for recycling from generic catering facilities and food production installations, where deep frying processes have been carried out. Usually these products are collected in authorised factories and processed simply (washing, filtration, drying, *etc.*) before being marketed for technical/industrial uses.

ANFA (Animal fats): products derived from the rendering process (sterilisation, cooking and melting of animal tissues). Only fats belonging to category 3, as classified according to EC Regulation 1774/2002, can be used, with some restrictions, for feeding purposes.

EBE (Oils extracted from exhausted bleaching earths): these oils are recovered, generally by means of hexane extraction, from the exhausted bleaching earth coming from the refining process of oils and fats. These oils are forbidden for food and feed uses and are generally employed for low-value technical applications.

FISH (Fish oils): these oils are obtained by rendering of whole low value fish or from food industry fish waste, such as canned tuna fish, smoked salmon, salted sardines, dried fish stock, *etc.*

HYBY (Hydrogenated byproducts): this category is covered by fully hydrogenated acid oils from chemical refining.

FACS (Fatty acids calcium soaps): solid product prepared by the neutralisation of fatty acids with calcium hydroxide (in direct or indirect ways).

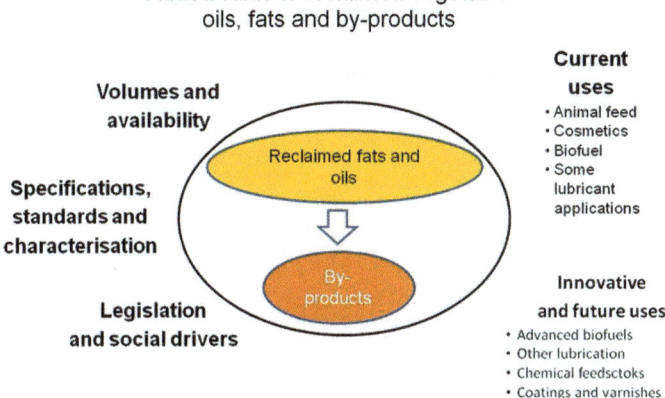

Figure 6.2   General scheme for the valorisation of reclaimed oils and fats.

A great deal of the mentioned by-products end up as animal feed agent in different forms. Animal by-products and catering frying oils and its production side streams (effluents water, sludges, *etc.*). The objective of this chapter is to evaluate and describe applications for some of these coproducts to be used in added-value products other than as feed. The main focus of the work is the study of current and potential technical applications for reclaimed oils and fats (including frying oils and animal fats animal/animal by-products) and also biodiesel coproducts such as glycerine. Physical and chemical properties are generally regarded as the main factor affecting the use of these materials in added-value applications. However, issues such as regulations, feasibility, quality, commercial supply and demand, and technical factors play a fundamental role in valorisation as shown in Figure 6.2.

6.2   Focus on Reclaimed Oil and Fats

Used cooking oil from various catering facilities and animal by-products cannot be recycled into the feed supply chain, and are widely considered to have restricted potential for valorisation (Figure 6.3).[8–10]

In the EU, catering waste and animal by-products are justifiably highly regulated, following the outbreak of the BSE in the late 1990s, restricting their valorisation potential to non-feed and non-pharma applications. Recent estimations of the amount of UCO annually generated worldwide have proven to be difficult but previous reports indicate 0.7–1 million tonnes/year are generated at EU[11] level, with 75–150 000 tonnes/year coming from the UK.[12] According to the University of Minnesota, the US alone produces approximately 1.5 million tonnes/year (including yellow and brown grease)[13] and reports estimate some 2–3 million tonnes/year of UCO are generated in China.[14] These figures would account for a total UCO generation worldwide of ca. 5 million tonnes/year. In addition, more than 12 m tonnes of rendered animal fat and grease are produced per annum worldwide (3 million tonnes per

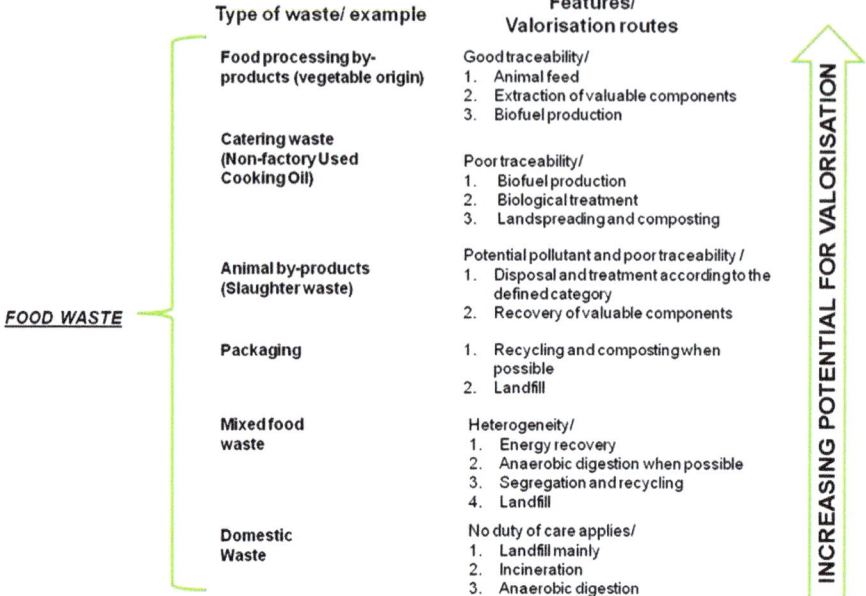

Figure 6.3 Classification of FW types, from Ref. 1a.

annum in the UK alone);[15] 2 million tonnes of coproducts from the biodiesel industry are generated *via* transesterification as biodiesel glycerine, where in many cases reclaimed oils and fats have been used as feedstock.

The use of these coproducts in technical applications are affected by issues such as quality, technical feasibility and regulations, for instance, the US does allow the use of UCOs in the food supply chain.[16a,b] Clearly social and economic issues will have an impact on the potential technical uses for such co-products.

6.2.1 Used Cooking Oil: Definitions

Current and potential applications for UCO include its utilisation in fuel boilers, lubricants/surfactant precursors and in biodiesel production.[17] UCO conversion to biodiesel is the most economical from the feedstock perspective; it can be sold at £550–750/tonne to large biofuel producers, as compared to the final product (biodiesel) which can be marketed at £1000/tonne.[18] FVO by contrast, generally contains lower levels of moisture, impurities and fatty acids and a high eluteable content, which may increase its applicability to feed (subject to traceability) in addition to other technical applications besides biodiesel.

6.2.2 Animal By-products and Rendered Fat

The meat, poultry and fish industries produce the largest quantities of waste within the food industry. Waste from these sources vary in type

Table 6.1 Description of animal by-products.

Category/risk	Definition/type of waste	Prescribed treatment
Category 1 (Very high risk)	Animal parts or animal types nonsuitable for human consumption (carcasses and BSE/other disease-infected materials)	Destruction is by incineration Recovery of valuable components for fuel applications, (rendered fat)[20]
Category 2 (High risk)	Dead animals Manure and digestive material Other material different to category 1 and 3.	Biological treatment Anaerobic digestion[21] Recovery of valuable inorganic and organic components[22]
Category 3 (Low risk)	Suitable for human consumption but discarded for commercial reasons (packaging, wrong transport, expiry date, *etc.*)	Recovery of valuable components for a range of applications: pet food, biofuel production and cosmetics

and composition but are generally highly polluting (*e.g.* blood, fats, and residues from intestines, partially digested grass or manure). In addition, industrial processing water is discharged as a liquid effluent that may have high nitrogen content or high levels of BOD/COD (biological and chemical oxygen demands, respectively). Various treatments for pathogens and pollutant removal are required for these streams, with common processes including thermal, biological or chemical treatments.[19] For some animal by-products, it may be challenging to develop routes towards recycling or reutilisation due to health and environmental concerns. The regulation on the use of animal derived byproducts is very strict in the EU, establishing three different categories which determine the end use of such coproducts (Table 6.1).

6.3 Characterisation of UCO and Triglyceride-Rich Food Waste

Frying oils mainly comprise triglycerides, monoglycerides, diglycerides and variable quantities of free fatty acids (FFA; 5–20% w/w), generated during the frying process.[23] Rendered fats contain variable amounts of FFA and are semi-liquid/solid at room temperature.

As with any material to be employed as feedstock its specification, physical and chemical properties are of key importance. The composition (free fatty acid content, moisture, impurities and fatty acid profile mainly) will determine the aptitude of various recovered fats and oils to be employed in various application.

Therefore understanding the specification of UCO and recovered fats is of critical importance alongside the need for standardisation, aiming for the recovered fats and oils to be employed as a consistent feedstock for various fuels and chemicals.

6.3.1 UCO Physical and Chemical Properties

The process of frying has a significant effect on the physical and chemical properties of the oil employed. These differences will be affected by the FFA profile of the parent oil, and by the period of frying undergone by these oils in addition to the type of process it has undergone.

As a general rule, oils recovered from frying processes will have, in comparison with the parent virgin oils:

- Higher viscosity, with fats being solid at ambient temperature.
- Higher free fatty acid content due to the hydrolysis of triglycerides plus oxidation promoted by the presence of water at high temperatures.
- Lower smoke point; An oil or fat subjected to thermal stress will decompose at a certain temperature (usually 150–180 °C) forming smoke that will be comprised of fatty acids, glycerol and decomposition products such as acrolein, hydrocarbons. This is known as the smoke point. Generally, the smoke point of oil increases as the free fatty acid content decreases and as the degree of refinement increases. A lower smoke point will be expected in oils that have been employed in long periods of frying. Decreasing the smoke point has a strong impact on the quality of food in frying processing. Also intermittent frying has a markedly greater effect on oil deterioration than continuous frying, therefore monitoring of oil frying usage is essential to ensure minimum standards of quality in food processing.[24]
- Higher amounts of moisture and impurities, as a result of the frying process, where ambient moisture and water contained in food are absorbed by the oil.
- Lower resistance to oxidation, resulting from the thermal degradation of natural antioxidants present in virgin oil, the hydrolysis processes promoted by the presence of water and the formation of auto-oxidative species. The stability of the oil can be improved by the use of additives.[25]
- Higher composition of polar component, such as polymers, aldehydes, short-chain fatty acids and ketones. These species are formed by the inter-action of fat/oil with air, food and moisture at high temperature conditions (150–180 °C). At these conditions, a number of chemical processes will occur (peroxide formation/oxidation, Maillard reactions, hydrolisis, *etc.*), leading to the formation of, amongst others, polymeric polar species responsible for the formation of sludge and viscosity increase in frying oils.

Attending to these parameters and drivers, certain recovered oils from frying will have better optimum properties for use in value-added applications.

6.3.2 Quality of UCOs

The quality of oil or fat used for frying is of paramount importance to ensure adequate standards for fried food. The standardisation of thermal rancidity and quality control is a challenging task, inflected by technical, cultural and regulatory issues. For instance, the threshold for quality control differs in different countries. The EU stipulates a polar content below 25 wt.% and a smoke point above 170 °C for oil to be usable for frying, when in the US the limit is given by an FFA value below 2 wt.%. In other countries such as Japan, Acid Value 2.5) and carbonyl value (50) is used to decide when UCO should be discarded. There are as many types of recovered oils and fats as there are processes where they are employed or recovered; therefore it is difficult to propose a specific classification for these materials.[26] A number of rapid analysis techniques (of viscosity primarily) have been proposed to evaluate shelf-life in order to establish the quality of a product.

In addition, the shelf-life and period of use for the oil introduce further variability to the quality of the remaining UCO, where the main driver for regulators and industry is to deliver acceptable standards for the use of vegetable oil in frying and food preparation. Generally, there is a direct relationship between the fatty acid profile, fatty-acid content and degree of refinement with the potential for degradation of the oil.[27]

The critical parameters to classify oils and fats according to its composition are;

- fatty acid profile;
- free fatty acid content;
- polar and eluteable content.

It is difficult to establish a general classification for UCO, frying oils and fats solely on the basis of composition. It makes more sense to produce generic specifications based on critical parameters such as viscosity, melting points and oxidative stability,[28] in order to facilitate the utilisation of recovered fats and oil as raw materials.

6.3.3 Analysis for UCO

The determination of specifications for UCOs and fats is important when determining the end application.

Frying-oil specifications are completely aligned with the need to ensure food quality when frying,[29] therefore any new processes or applications utilising UCO and fats as a raw material will have to adapt to the variable quality of the available material. The clearest example of this is the production of biodiesel, where preprocessing (drying, deodorisation, esterification, *etc.*) and adequate downstream processing (additivation, distillation, *etc.*) are required to deliver a quality product.

Therefore the key driver to monitor quality and shelf-life of UCOs and fats for both the food and biofuel industry are the following:

- Moisture and impurities, generally determined by centrifuging aided by titrimetric techniques such as volumetric Karl Fischer (moisture titration).[30]

- Determination of acidity, which is a key parameter for the usability of UCO. There are several methods used to determine the acidity, either using wet acid–base titration with KOH or NaOH classical titration[31] or potentiometric methods[32] as common methodologies. The latter has the advantage of determining endpoints more accurately than the classical methods, making it nondependent on colour change.
- Viscosity, which is measured by a number of rheological techniques. Measuring viscosity can be an inexpensive and simple method to determine the shelf-life and degree of deterioration/ quality of the oil in comparison with other highly technical methods.[33] Viscosity is especially important for the use of UCO and others in lubrication and applications involving the mechanical performance.
- Free fatty acid profile and saponifiable materials/eluteables: to quality, purity and aptitude for an oil/fat to be employed in certain application. A common technique employed is Gas Chromatography (GC), widely used both in feed and technical uses such as biodiesel (Figure 6.4).[34]
- Polar component: Total polar material (TPM) evaluation in oils and fats may be another good indication of the degree of degradation of the product. Some polar components are naturally present in fats and oils (tocopherols, sterols, glycosides and phospoholipds primarily), whereas others such as oxidised triglycerides and fatty acids and their polymers form during frying and exposure to air. Methods such as size exclusion chromatography, TLC and others can be employed to elucidate this analysis both qualitatively and quantitatively.[35]
- Iodine value (IV), related to the degree of unsaturation in the oil. This is defined as the amount in grams of iodine reacting with 100 g of oil/fat, being halogens readily reactive with double bonds in the side chain from the fatty acids in triglycerides. The iodine value is a parameter that correlates both with oxidative stability and other properties such as viscosity and lubricity.
- Oxidative stability, consisting generally of the evaluation of the performance of the material under accelerated thermal oxidation of oils/ fats in the presence of air (Rancimat method).[36] Resistance to stability generally will increase for fats and oils with decreasing IV, and therefore with the right balance of monounsaturated and saturates present in addition to the presence of natural or added antioxidants in the oil,[37] and will decrease with the length of use of the oil.[38]

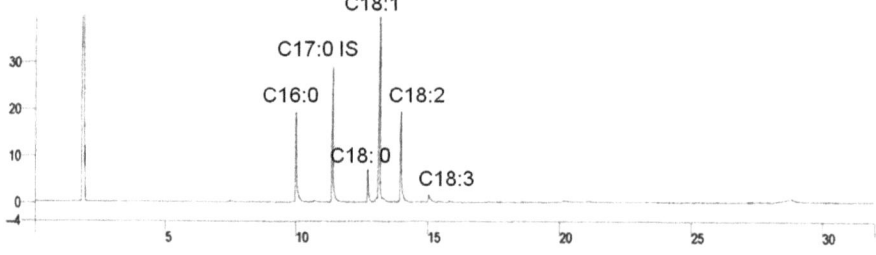

Figure 6.4   Typical GC free fatty acid profile for an UCO.

The oxidative stability of vegetable oils is determined in a complex way by their fatty-acid composition and tocopherol contents. Highly unsaturated fatty acids are more susceptible to oxidation, however this susceptibility is affected by their position within the triacylglycerol molecules, with the intermediate position (sn-1) being slightly protected. Among the tocopherols, a-tocopherol is the most efficient in scavenging peroxyl radicals, but its antioxidant activity in vegetable oils is partially minimized because of participation in side reactions leading to increased rates of initiation. With the identification of efficient synergists, the stability of vegetable oils might be improved.[39]

- Other techniques: Near and Mid-IR are reliable techniques suitable for the determination of triglyceride content in foodstuffs and also to support the determination of frying-oil quality parameters such as viscosity, smoke point, FFA and TPM values. In addition, HS-GC may be employed in order to determine the nature of the volatile components, HPLC and GC-MS may also be used for the determination of the polar components. NMR could be used for the characterisation of the mentioned parameter, although it is also employed for the quatification of oil and solid fat content (SFC) in seeds and foodstuff.

6.4 Legislation and Social Issues Regarding the Valorisation of UCO and Triglyceride-Rich Food Waste

Historically, UCOs from catering and food production facilities have been recycled into the human food supply chain as an animal feed supplement. However, regulation linked to the outbreak of the BSE in the late 1990s, has restricted the use of UCO as a component for animal feed. UCOs are these days regulated under the animal by-products regulations.

More recently, the contamination with pesticides and mineral oils of some materials intended to be recycled into the human food supply chain (FVO) has increased pressure and restrictions for these types of materials. FVO is indeed used as an animal feed supplement to add calorific/nutritive content to pellets used for poultry, pork and beef breeding, with the exception of those processing facilities where meat has been processed. In recent years there have been attempts within the feeds and fats industry to deregulate certain types of materials with a high degree of traceability and removal of substances of concern (generally muscle fibre), allowing the recyclability into the food supply chain of highly traceable oils which have been in contact with meat.

UCO and other recovered triglycerides have different statuses as waste, and this status is either as waste or as product. Therefore the valorisation of which will be subjected to regulatory constraints stated in the waste framework directive (WFD) safegarding human and environmental health for the food supply chain in addition to target waste minimisation and proper treatment (either by recovery, recycling and other treatments) of food wastes (Figure 6.5),

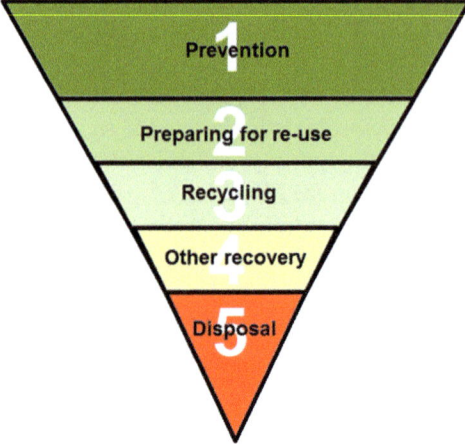

Figure 6.5   Waste hierarchy.

(mainly through disposal in landfill, composting, incineration or use in animal feed if feasible) as preferred routes for these materials as shown in Figure 6.4.

Therefore, the WFDs will contemplate the conversion of food waste such as UCO into a different shape or form such as a useful chemical product as fuel if:

- in the process the waste processor can prove the waste is no longer waste and has undergone a recycling or recovery process;
- an improved environmental situation has been generated through the processing of that waste by a suggested treatment route.

In this context, the Commission concluded that composting and anaerobic digestion (AD) offered the most promising environmental and economic results for bio-wastes that cannot be prevented.[40,41] As for oils and fats the best route would be the conversion into biodiesel or as heating fuel.

Here, UCO and recovered fats face a number of constraints when being employed in added-value uses, *e.g.* chemical products:

- The need to comply with established procedures leading to a Quality Protocol for the product ultimately produced.
- The potential disincentive of what REACH may become in relation to the waste status of UCO and some animal fats for their conversion into chemicals.

6.4.1   Quality Protocol

The quality protocol (QP) is a requisite established for certain types of waste that determines when a "waste has ceased to be a waste" and therefore has fulfilled "end-of-waste" criteria as having undergone "recycling". So far, the only QP suggested for oils and fats where the material has been "recycled" is for the biodiesel process when the material is burnt in the motor engine.

This procedure requires compliance with an "approved standard", which in the case of the conversion of UCOs and some recovered fats into chemicals and other potentially valuable products (materials and other fuels) does not exist, due to a number of reasons mainly related to the potential variability and lack of traceability of the raw materials. The reprocessing of UCOs and animal by-products into different products regardless of their purpose, with or without the production of energy may constitute "recycling" or "other" recovery or possibly even pre-treatment steps. At that stage, every case would have to be analysed individually, in order to evaluate whether the process in question has a positive environmental effect or having undergone recycling. These procedures may be perceived by potential producers and organisations both as an opportunity to acquire end-of-waste status or an administrative constraint.

The quality protocol/"end-of-waste" approach is intended to incentivise processors generating waste, to address their issues with food waste and prove the viability of an end-of-waste status by producing as much information as possible for the waste in question. This approach is expected to have very significant commercial implications with the extension of the Quality Protocols approach, being a positive step towards strengthening the economic potential of wastes where there is uncertainty over end-of-waste status.

6.4.2 REACH Legislation

A potential policy and regulatory disincentive to the reprocessing of food wastes into chemical substances is the consequence of end-of-waste status and chemical substances legislation, most notably through Regulation (EC) No. 2006/1907 on the Registration Evaluation Authorisation and Restriction of Chemicals (REACH).[42] This requires all those who manufacture in the EU in quantities of one tonne or more per annum per manufacturer (or who meet the threshold criteria in an importer capacity) to obtain a registration for the chemical concerned, without which the substance cannot be placed on the market within the EU. Despite provisions that enable producers/ importers to share the cost of obtaining all the necessary hazard and risk data required to register the same substance, the testing and administrative costs of achieving a registration are nonetheless considerable. Thus, small-scale producers in the EU, particularly of novel substances/ mixtures resulting from food waste reprocessing, may ultimately find the compliance costs of REACH legislation a major barrier to the commercial viability of the process. With other major economies outside the EU also showing interest in adopting similar legislation to REACH (including the US and China),[43] even manufacturing and distribution outside of the EU may at some point become unviable and therefore the usability of this type of as feedstock within the manufacturing industry.

6.5 Valorisation Routes

The situation in the EU with regard to FW/ Bio waste/ FSCW regulation exemplifies the potential impact that research in the area can have into driving policy changes to accelerate the valorisation of FSCW beyond current practice (animal feed or energy recovery, *etc.*). This area is not free of challenges, since "grey areas" such as the distinction between waste and by-products, strong political drives for the diversion of FW into AD and policies such as "end-of-waste" regulation and REACH will have an impact on future practise.

There is therefore increasing pressure to find sustainable routes for the valorisation of UCOs and reclaimed fats. An overview of the different options is shown below:

- **Production of biodiesel**, which is the most widely employed method for the valorisation of fats and UCOs. As previously mentioned, in the UK a Quality Protocol criteria is in place for the production of biofuel, the only one currently existing for recovered oil and fats derivatives. (Reference: Biodiesel, Quality Protocol – End of Waste Criteria for the production and use of biodiesel derived from waste cooking oil and rendered animal fat (quality biodiesel), Environment Agency for England and Wales and the Northern Ireland Environment Agency, 1 June 2009. Biodiesel can be employed in fuel for car consumption or for heat electricity generation,[44] where blends of UCO-biofuel with mineral diesel can also be used. The quality of this product in relation to the one derived from virgin feedstocks has been widely discussed in literature and the general consensus is that the UCO-derived biodiesel has similar properties to conventional biodiesel, with the exception of its resistance to oxidation/ and cold-flow performance, which is generally addressed by the use of additives.
- **Energy recovery**, use as fuel for industrial boilers. Suitable for all types of UCOs including blends of biofuel coproducts, namely oleins and free fatty acid phases.
- **Use of feedstock for anaerobic digestion**, limited to some types of processes, since the detrimental effects of oils and fats in "conventional" AD processes are also well known. At a national (UK) policy level considerable emphasis is already being placed on AD as the favoured waste-management option for bio-wastes going forward.
- **Use as pet food**, only suitable for rendered fats derived from category 3 animal by-products. The product must meet customer specifications.
- **Recyclability into technical applications**, either by use as a net product or by chemical modification. This is *a priori* the most attractive approach from a research and added-value approach. However, classification of such materials as waste, may be difficult in their conversion into chemicals, particularly in the EU, where regulations such as REACH, demand much bureaucratic effort. In order to displace AD from its apparent "preferred status" for the recycling of oil sound economic and environmental advantages will need to be proven.

6.5.1 Use as Fuel; Production of Biodiesel

The use of virgin oils for biofuel production has generated significant controversy, deemed the "food versus fuel issue". In this regard, recovered oils and fats as feedstock for biodiesel can offer an appealing alternative to achieve more sustainable biodiesel production worldwide,[45] avoiding the use of virgin food crops in fuel applications. However, the potential for biodiesel production from UCO remains limited, as UCO valorisation can only meet less than 30% of the world's biodiesel demand, with the remaining amounts to be sourced from other feedstocks.

Waste oils can be effectively converted, *via* transesterification with methanol or ethanol, into fatty acid methyl esters (FAME; biodiesel) (Scheme 6.1 and Scheme 6.2) using a range of catalysts including solid acids and bases. UCO a much larger free fatty acid (FFA) and water content than those of virgin oils, both of which are detrimental in FAME production.[46] The most widely extended biodiesel production process from UCO is the homogeneous base-catalysed transesterification of glycerides, which requires a pre-treatment of the FFA with $MeOH/H_2SO_4$ (Figure 6.6), followed by the recovery of FFA/unrefined FAME (FA material) and unreacted methanol.

There are also literature reports on the use of heterogeneous catalysts that can efficiently catalyse the simultaneous esterification of the FFA as well as the transesterification of the triglycerides present in the waste oils.[47,48] Heterogeneous catalysts such as alkaline oxides,[49] supported enzymes and carbonaceous materials have also recently received increasing attention due to their more environmentally sound credentials, compared to their homogeneous equivalents.[50] The major issues with the utilisation of heterogeneous catalysts for biodiesel production from UCO are linked to lower conversion rates, deactivation of active sites due to the presence of FFA and moisture and reusability issues, which is one of the major drawbacks of the most widely utilised heterogeneous catalyst for biodiesel production (CaO).

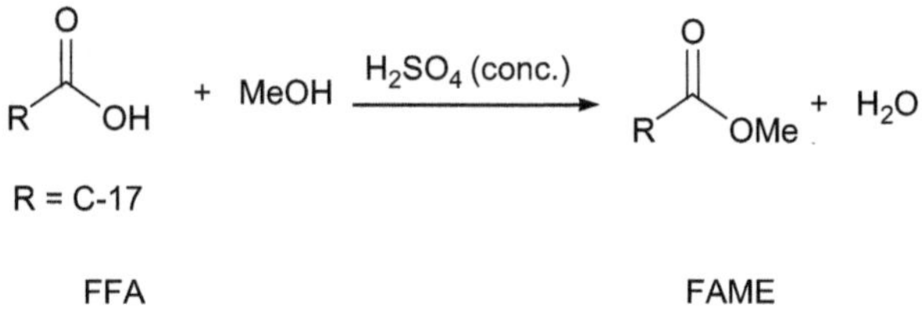

Scheme 6.1 Esterification reaction for acid pretreatment.

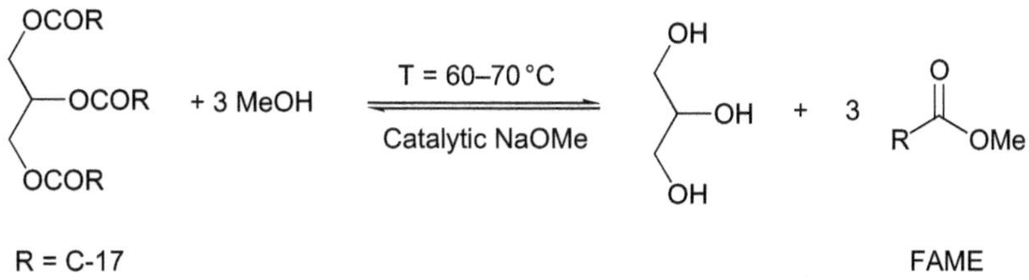

Scheme 6.2 Transesterification reaction of triglycerides.

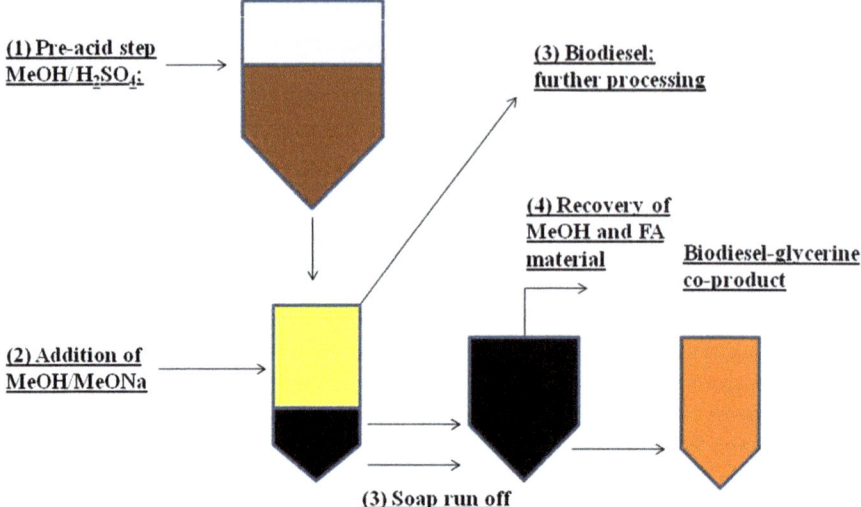

Figure 6.6   Biodiesel production from UCO.

An interesting example reported in literature obtained CaO from FW (*e.g.* eggshell and mollusc shells by calcination) which could catalyse the production of FAME from pre-esterified UCOs into biodiesel.[51] The use of an inexpensive source for the production of an active catalyst can improve the cost competitiveness of the process. Enzymatic conversion has also been successfully evaluated, showing great potential for the development of a commercially available lipase-based process for the conversion into biodiesel of high FFA content UCOs.[52]

Employed feedstocks include waste frying, olive, rapeseed or sunflower oils, and rendered animal fats as well as others sourced from food industries,[53] restaurants and catering facilities that are largely exposed to air, high temperatures and moisture,[54] all parameters that increase FFA content in oil.[55]

In summary, UCO is currently one of the most attractive feedstocks for the production of biodiesel. Its lower market cost in comparison with virgin oils and recovered factory oil, favourable government incentives,[56] plus the possibility to recycle waste from the catering sectors, make UCO-derived FAME a commercially and environmentally acceptable feedstock for biofuel production.[57]

A relevant development in the area showing the potential for the valorisation of UCO and other co-products (tallow, yellow grease, *etc.*) is exemplified by North Cave Ltd. (Figure 6.7), a new company formed as a joint venture by Brocklesby Ltd. and Greenergy Biofuels. The company will be aiming to process ca. 50 ktonne of high-FFA UCOs and tallow blends *via* acid esterification as pre-treatment for the production of biodiesel by mid-2013. This new company will allow Greenergy to become the largest biofuel producer from waste products in the world.

Figure 6.7 North Cave Limited.

6.5.2 Focus on Recovery of Fats and Oils from Food Waste for Biodiesel Production

Traditional uses for tallow and lard included fuel, cosmetic precursors and components of pet-food supplement. Due to changing regulations, biofuel production from rendered fats has recently increased in importance in the EU and the UK.[66] Brocklesby Ltd. has developed a process that recovers oil and fats from miscellaneous oil/fat-rich food waste (Figure 6.7), including animal by-products employed in biodiesel production with up to 30% triglyceride content. The process currently in operation has the potential to process 2000 metric tonnes per year of FW sources (solid waste, triglyceride rich waste, processed as Category 3 animal byproducts). The recovery of triglycerides is carried out under a process largely based on wet rendering technologies,[67] with high efficiencies (up to 98% yields). The novelty of the process is the flexibility of the approach where a number of products can be processed due to the possibility of working under wet and semi-solid conditions. Despite the high process efficiency, a starchy/fibrous coproduct is sent offsite for composting. The company is currently working on the reduction of the waste volumes through various alternative treatments including the use of microwave heating, with significant potential to generate a solid fuel precursor from the wet co-product. The extraction process involves particle reduction as a first stage by means of mincing, followed by thermal treatment at temperatures below 100 °C, in order to break down the fibrous/ lipid cells, freeing the melted oil adsorbed/present inside the cells. This is followed by a mechanical separation

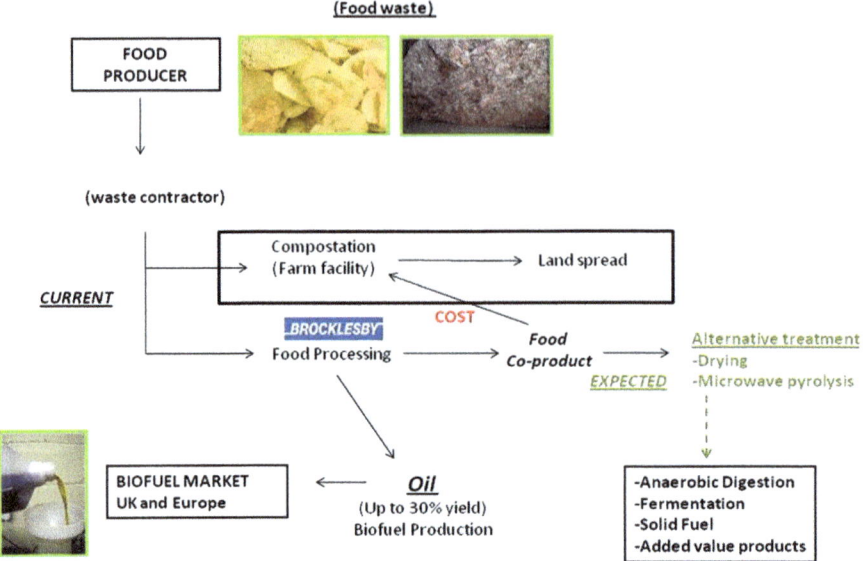

Figure 6.8   Business model scheme for food waste processing.

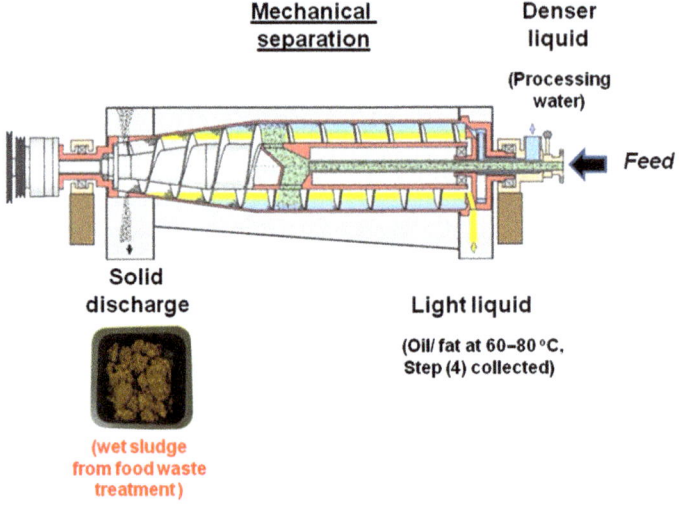

Figure 6.9   Centrifuging step for oil recovery.

via a tricanter centrifuge (Figures 6.8 and 6.9) where the oil/fat is separated from the solids (wet sludge) and the processing water.

Applications for the wet residue are currently being investigated, with possible uses as substrate for fermentations and feedstocks for microwave treatment. The latter has the potential to produce bio-oil and char.[68] Preliminary work looked into the possibility of combining an effective way of

Figure 6.10 Compacted waste sludge.

drying waste from oil extraction (Figure 6.10) with a simultaneous metho-dology to obtain bio-oil and a high calorific value biochar. For instance, the gross calorific value found in the char generated from microwave treatment was 27.4 kJ/g, 22% higher than that of the dried solid obtained from the parent material. The main components found in the bio-oil were levoglucosan and FFAs, showing the potential of microwave pyrolysis as a suitable treatment both for oil extraction, homogenisation and conversion of FW into added-value chemicals.

6.5.3 Alternative Biofuels: Hydrocracking Applications

Apart from biodiesel-related research, there are some other interesting catalytic routes to biofuels from waste oils. One of which is the production of biohy-drogenated diesel.[58] These routes involve the treatment of oils with hydrogen using heterogeneous catalysts (typically Ni-Mo and/or Co-Mo and related desulfurised catalysts) as well as solid acid catalysts such as H-ZSM-5 or sulfated zirconia to generate a paraffin mixture upon removal of oxygen and hydrogenation of all C=C double bonds.

The resulting product obtained from low-quality feedstocks was reported to be gasoline (>30%) and gas oil (>30%) when using heterogeneous catalysis zeolites (HZSM-5) and sulfated zirconias.[59–61] Optimum yields of gasoline are obtained between 400–430 °C, at a pressure of 10 bars of hydrogen and 90 min of reaction in a batch reactor.

This route is particularly interesting as glycerin is not generated as a coproduct, with all the triglycerides and FFA being converted into hydro-carbons. Biohydrogenated diesel has been proposed as the next-generation biodiesel and has been commercially demonstrated by Neste Oil.[62,63] A recent study from Norwegian researchers has reviewed and compared environmental impact, life-cycle assessments and costs between different biodiesel fuels (transesterified lipids, hydrotreated vegetable oils and woody biomass-to-liquid Fischer–Tropsch diesel) and come to the conclusion that hydrotreated oils from waste or by-products including tall oil, tallow or waste cooking oils outperform any other diesel product in terms of environmental LCA impact and costs.[64]

In Greece, Bezergianni *et al.* have recently described how the production of hydrogenated green diesel from waste oils can take place in high yields (>95%) using typical commercially sulfided Ni-Mo catalysts.[61] Low-pressure values generally maximise biohydrogenated diesel production as opposed to high temperatures that favour cracking reactions. In all cases, the heteroatom removal in the systems was quite efficient, exceeding 99% for sulfur and nitrogen and over 90% for oxygen.

The main drawback of this methodology is the use of hydrogen in systems that generally require high pressures. Unless it can be obtained from a renewable resource (*e.g. via* aqueous phase reforming of another waste residue/renewable feedstock), the proposed approach cannot be commercially feasible. In this regard, coupling an APR approach with oil hydrotreatment processes could be the ideal sustainable solution for the valorisation of waste oils.

6.5.4 Beyond Fuels, Use of Recovered Triglycerides as Chemical Feedstocks

Biodiesel and the use of UCO as a fuel precursor, are the most common valorisation routes for UCO, however cost, availability and the properties of this food coproduct and other recovered triglycerides have facilitated the development of alternative technical uses to the transformation into biofuel in many industrialised countries (Figure 6.11).

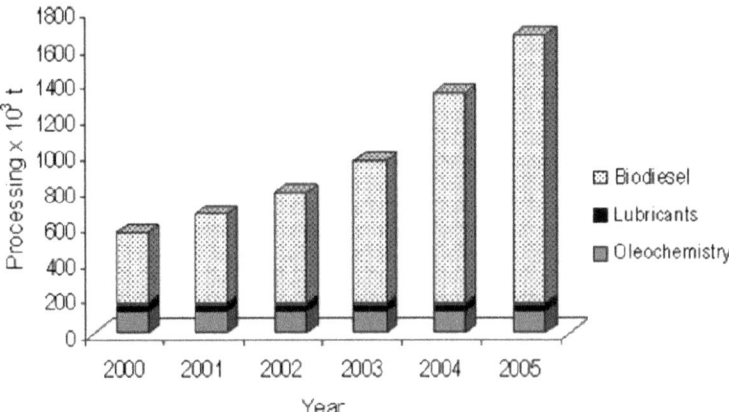

Figure 6.11 Volumes of lipid-derived technical products in Germany.[69]

For instance, products such as surfactants, lubricants, coatings, polymers have been generated from either food or nonfood triglycerides and fatty acids. Selected examples reported the use of vegetable oils and fats in lubrication,[70] paints and drying applications,[71] which can potentially be economically attractive and sustainable "end-of-waste" routes for UCO and recovered fats.[72]

6.5.4.1 Vegetable Oil, Fats and UCOs as Lubricants

Biorenewable, nonhazardous and biodegradable components for lubrication are expected to considerably increase in future years, driven by environmental and health concerns (toxicity, not being readily biodegradable, *etc.*) as well as by the large demand of mineral-based lubricants worldwide (37.5 million tonnes/year).[73]

Vegetable oils, greases and fats either neat, in blends, as well as chemically modified, or with additives have become a promising alternative to mineral- and synthetic-based oils with a large number of examples reported in literature.[74] The preferred applications for vegetable oils in lubrications are those in which thermal/oxidative stability is not a critical issue. These include cutting fluids,[75] low working temperature hydraulic fluids[76] and others. Although the reported examples for UCO as a lubricant are limited[77] due to their physicochemical properties particularly related to thermal and oxidative stabilities.

However, additives are required to improve oxidative stability and viscosity at high and low temperatures to achieve a similar performance to those of their synthetic/mineral equivalents. These include anti-wearing agents (zinc dialkyldithiophosphates: ZDDPs), antioxidants (hindered phenols, aromatic amines and others), detergent/dispersants (sulfonates, salicylates and others) *etc.*) chemical modification,[78] (*i.e.* improvements in thermal/oxidative stability, viscosity modification boundary lubrication, *etc.*), deodorisation[79] (*via* distillation to improve flash point properties) and/or blending with mineral products to meet the specifications.[80]

Research in this area is needed to achieve further developments. For instance, the free fatty acids available in large amounts of acid oils/UCOs could be rectified by conversion into soaps, amides (*i.e.* partial emulsification of water in cutting fluids), fatty acid alkyl esters (FAAE) or neopentyl triol esters (resulting in increased thermal/oxidative stability in hydrolic fluids).[81] Other modifications likely to improve properties such as viscosity index or oxidative stability are conversion into estolides,[82] epoxidation and partial oxidation (Scheme 6.3).

In summary, UCOs and derivatives have significant potential to become environmentally friendly alternatives to mineral-based lubricants, with low volatility, high viscosity index, high biodegradability and lower production costs. The utilisation of UCO in this area is, however, not free of challenges, due to variability of composition and lower oxidative and thermal stability in comparison to currently available virgin oils and mineral products.

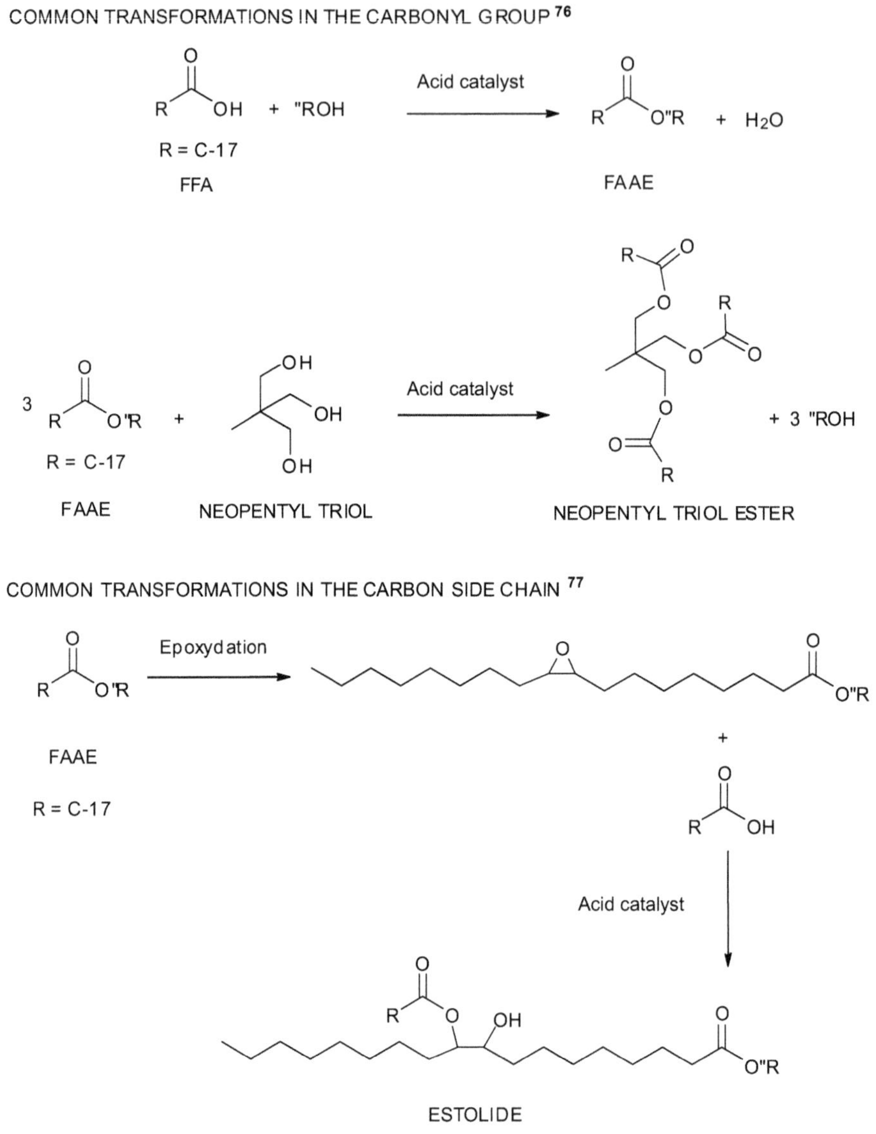

COMMON TRANSFORMATIONS IN THE CARBONYL GROUP [76]

R = C-17

FFA

FAAE

3

R = C-17

FAAE   NEOPENTYL TRIOL   NEOPENTYL TRIOL ESTER

COMMON TRANSFORMATIONS IN THE CARBON SIDE CHAIN [77]

FAAE

R = C-17

ESTOLIDE

Scheme 6.3   Useful chemical transformation for UCOs.

6.5.4.2   *Mould Release Agents*

A lubrication application where the use of UCOs would be preferred is precast concrete production. This term refers to the concrete unit, structure or member that is cast and cured at the factory and subsequently transported to the building site where it will be used. This process requires releasing agents that prevent the concrete bonding to the mould after curing.

Moulding release agents will require to be inert to the curing reaction, low cost and able to be applied without major difficulties, furthermore they should not produce uneven colouring on the concrete surface.

The concrete industry has traditionally employed large tonnages of mineral-based emulsions as release agents primarily derived from the motor industry; however, directives make the use of sustainable moulding agents compulsory. By using vegetable oil derived products this industry may overcome the current challenges it faces.[83]

6.5.4.3 Cutting Metal Fluids

UCO has been tested as a suitable raw material for conversion into metal working fluids, as a renewable-based alternative to petrochemical sources. Liquid lubricants are constituted by base oil (main component) and a number of other additives with different functionality (antiwearing, stabilisers, antioxidants, detergents, *etc.*), changing depending on the application.

The use of triglycerides in lubrication and cutting working fluids might be done by using the net triglyceride or through chemical modifications such as partial conversion into soap, amides or transesterification into alkyl ester to improve/ modify the properties of the formulation (*i.e.* performance, partial emulsification of water in cutting fluids, *etc.*).

High oleic sunflower oil-UCO (HOSO) (iodine value 75–95) is a vegetable oil with a large number of technical applications due to its good behaviour of oxidation and thermal stability plus adequate viscosity and high-temperature performances, which makes it a good candidate to become the base oil (main component of a lubricant) for a bio-lubricant either net or as a blend with conventional mineral-based oils.

6.5.4.4 Cosmetic, Paints and Varnishes

UCOs have also been proposed for the production of bioderived coatings, paints and varnishes. These industries are moving towards safer (*i.e.* nonflammable, non-VOC containing) alternatives as the mineral base of these products is a large source of contaminants. Most varnishes are a blend of resin, drying oil and volatile solvent, from which alkyds are the most important class of resin in the coatings industry.

Alkyds are made of an alcohol such as glycerol, a dibasic acid and oils including castor, coconut, linseed or soybean. Vegetable oils with a high level of unsaturated fatty acids have been tested as varnish agents (*i.e.* linseed, soy bean, high linoleum sunflower and castor oil), by chemical transformations under controlled mechanisms of polymerisation.[84] Condensation with polyols for the formation of alkyd resins, radical polymerisation and oxidation are required to form a solid polymerised matrix from UCOs. The main challenge with the use of these types of UCOs within these applications is their relatively short shelf-life and poor oxidative stability. Oils degrade upon frying leading to an increase in the FFA content and polymerisation reactions, which could be overcome by the introduction of functionalities in the carbon chain of the parent oil (epoxydation,[85] ozonolysis/ introduction of carboxylic groups to increase hydrophilic properties of paints, insertion of olefins, *etc.*).[86]

6.5.4.5 Fabric Softeners

Various applications have also been found for the use of recovered oils and fabric softeners. Recovered rapeseed and palm oil can be used. The oil is converted either into amides or ethoxylates and produces a high-quality fabric softener.

6.6 Soapstocks; By-products from Oil Refining, Recovered Triglycerides and UCO-derived Biodiesel

6.6.1 Introduction

Edible oil purification, biodiesel production, AOCHE production and pulp milling processing all generate coproducts of variable composition that are generally regarded as soaps or soapstocks.[87] These result from a combination of various streams in an oil refinery (caustic neutralisation, transesterification, effluent disposal, deodorising)[88] and require further processing in order to recover valuable components. Some of these components are free fatty acids, triglycerides, nonsaponifiable matter,[89] different grades of glycerine,[90] and tall oils in the case of paper-pulp refining. Therefore soapstocks can be considered both as pre-consumer and post-consumer FSCW/FW, for oil refining and UCO-biodiesel processing respectively.[91]

Crude oils from the seed crushing/ purification process, for instance, contain variable amounts of gums, colour-impairing bodies and fatty acids.[92] In one of the steps in the purification process, oil is treated with a caustic solution in order to achieve fatty acid removal. During the process, fatty acids are converted into soaps and part of the neutral oil, glycerides and phospholipids, are also saponified, generating soapstocks with a relatively high carboxylate content that can form stable emulsions.[93] These soapstocks from oil refining are generated at a rate of around 6% to the oil produced,[94] where in biodiesel production from UCO soap produced can be up to 45 wt% to the feed.[95] There are different soapstock treatment methods, for instance the use of mineral acid and/or solvent treatment followed by phase separation as shown in Figure 6.12 which has been a common approach in industry. This method is relevant to the current work since it resembles the approach taken in the biodiesel soap treatment for fatty material and glycerine recovery (Figure 6.13).[96]

The end uses of the fractions recovered from various soapstock refining processes are diverse. Use as an animal feed component[97] is the most common, however, no less important are their utilisation as biofuel precursors,[98] in the

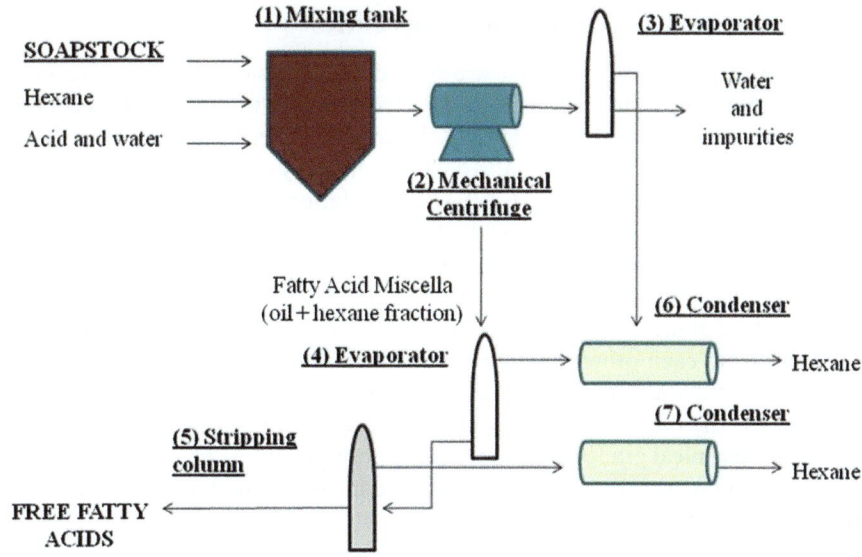

Figure 6.12 Classical soapstock treatment for free fatty acid recovery.

Figure 6.13 Scheme of the transformation of soap from biodiesel production.

production of tall oils[99] and the manufacture of chemicals from glycerine. Despite these higher-value applications, recovery is not always possible and disposal *via* waste treatment plants is also common.

When potential applications are being considered, the required grade and type of soapstock will have an impact on its uses. Feasible applications will therefore depend on the source of the co-product and subsequent treatment of the soapstock. The understanding of the nature of soapstocks and further purification methods for glycerine[100] and fatty acids will therefore be a key factor in developing suitable protocols for generating biodiesel glycerin with commercial value.

6.6.2   Biodiesel Glycerine (I): Generalities

Frequently, glycerol and glycerine are regarded as the same entity; however, it would be more appropriate to talk about glycerine as a mixture of glycerol and other components, primarily water.[101] Glycerine of different grades is produced from the processing of certain soapstocks, (commonly from production of fatty acids and of biodiesel (Figure 6.12) *via* acidulation and further purification (distillation, ultrafiltration, removal of salts and impurities[198] to the desired grade. Glycerine is employed in numerous industrial applications, taking advantage of the properties and chemical nature of glycerol.

There are a number of generally accepted commercial grades of glycerine that differ from each other in value and manner of performance, depending on the glycerol and water content, source and type of impurities. A general classification for the commercial grades of glycerine is summarised below;

Crude or technical grades: mixture usually above 60 wt.% of glycerol content, where certain types of biodiesel-derived glycerine (also known crude or effluent glycerine) will be classified (Figure 6.14).

Refined grades: These are commercial grades sold at least at 96–99.5 wt.% of glycerol content, being a popular specification within pharmaceutical and food markets, for instance into USP (U.S. Pharmacopoeia), and its European equivalent PH.EUR (European Pharmacopoeia II).

The presence of impurities in biodiesel-derived glycerine (methanol, salts, fatty acids and FAME primarily) and the classification of UCO-derived glycerine as a waste product[102] limits the potential of UCO-derived glycerine for purification into refined grades.

6.6.3   Glycerine: General Market Perspective

Glycerine markets have undergone significant changes over recent years. Glycerine is regarded as a volatile commodity in terms of its inconsistent

Figure 6.14   From left to right: UCO-derived soapstock, biodiesel glycerine(effluent), technical and refined grade glycerine.
Sourced from Brocklesby Ltd.

availability. This "volatility" is determined by factors including the supply and demand of several commodities including soap (carboxylates), free fatty acids, fatty alcohols, oleochemicals and FAME, the availability and cost of raw materials, weather conditions, regulatory costs, plant shut-down and politics. The glycerine market is small by chemical industry standards: approximately 1m tonnes for 2011, less than £2 billion, and forecast to increase to £4.4 billion by 2015. This factor is critical in order to understand how small over-supplies might affect this market such as increasing amounts generated by the production biodiesel (ca. 1.6 m tonnes).[104]

Most of the glycerine produced globally, as a by-product from fatty acids/soap production ends up in oral care and food products (Figure 6.15), nevertheless an estimated 160 000 tonnes are employed in different applications (plasticizer, paints, humectants, manufacture of alkyd resins, polyols, polyethers and other chemicals, *etc.*).[105] Prices of glycerol have experienced drastic changes over the last decade, caused by the large amounts of biodiesel glycerine that have come on to the market (above 1m tonnes in 2009),[106] dropping from 1000–1300€/tonne in the period 2000 to 2003 to 500–700€/tonne, between 2004–2010.[18] This market volatility has earned glycerine the reputation of being an unreliable commodity with producers seeking alternatives in certain formulations. In addition, various factors such as changes in raw-material costs for biodiesel, the latest economic recession or modifications in the biofuel incentive policies around the EU, have considerably affected the biodiesel market worldwide, causing biodiesel plant shutdowns.[3,107] These changes have caused prices of crude glycerol to fluctuate dramatically due to shortages to certain producers relying on crude biodiesel glycerine. Given this context, processes such as the purification of crude glycerol from biodiesel might be appealing,[108] however, the high capital investment required for this activity (distillation and other processes) and the regulatory implications of using UCO-biodiesel glycerine, restrict the access to high-quality glycerine markets such as USP glycerine. Biodiesel glycerine finds more attractive outlets in upgrading for technical applications and in its conversion into more valuable chemicals.[109]

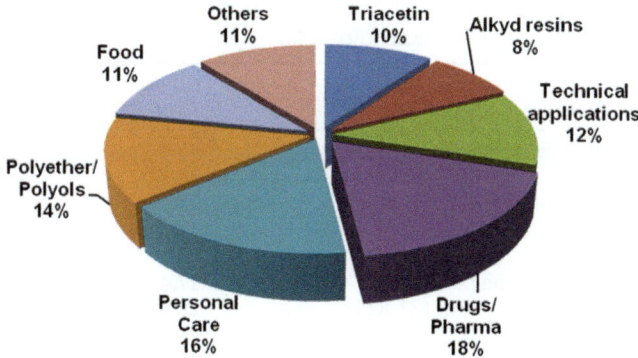

Figure 6.15 Different uses of glycerol by market volumes.[103]

6.6.4 Biodiesel Glycerine Applications

Biodiesel glycerine, as a catering oil by-product, has found treatment routes as a landspread agent; although environmentally acceptable, this is a costly activity. In the UK, the type of glycerine most suitable for this purpose must contain less than 3 wt% methanol (above that the concentration of methanol glycerine is considered a toxic and flammable substance)[110] and preferably potassium phosphates as the major inorganic component (formed after the use of MeOK in the transesterification step and phosphoric acid in the purification of biodiesel).[111] Alternatives to this disposal route are the objective of our research. Despite the limitations for this type of coproduct, there are a number of possibilities that have attracted interest recently.

6.6.4.1 Conversion into Technical-Grade Glycerine

Methanol recovery after the transesterification step is economically attractive for biodiesel producers. It reutilises one of the reagents to be employed as feed in the process (methanol costs £250–375/tonne),[112] whilst increasing the glycerol content in the glycerine. Methanol recovery should ideally be effective to the point of leaving less than 3% methanol in crude glycerine. Slightly acidic/neutral pHs (5–7) are preferred in this material as it increases the commercial value of glycerine. Distillation under vacuum to recover methanol and ultrafiltration[108] for certain types of applications (cosmetics, humectants, plasticiser, or chemical conversion amongst others) would both lead to technical-grade glycerine.

6.6.4.2 Energy Generation

Despite its lower calorific value compared with other fuels and biofuels, (Table 6.2) glycerine has been employed in energy generation.[113] For instance, Aquafuel Research (a UK-based company) claims to have developed a technology wherein glycerine from biodiesel production is directly employed in

Table 6.2 Calorific values for some conventional fuels and biofuels.

Fuel type	Calorific Value (MJ/kg)
Petrodiesel	46
Coal	35–42
Natural gas	38.1
Biodiesel (FAME)	37.3
Ethanol	30
Pyrolysis char	27.4
Biodiesel glycerol (37 wt% glycerol; 21 wt% methanol).	25.30
Pyrolysis oil	16–23
Wood (15 wt% water)	19.0

CHP (combined heat and power) diesel engines.[114] However, the glycerine still needs to be purified *via* distillation. Glycerine can be converted into biogas through anaerobic digestion. Biogas is effectively a mixture of methane, CO_2 and water plus impurities of differing proportions.[115] The optimization of this process can lead to the production of methane and/or hydrogen.[116] Methane can be burned in CHP units or supplied to the gas grid, making anaerobic digestion an attractive possibility for many biodiesel producers across the EU in order to effectively valorize biodiesel glycerine.[117] Anaerobic digestion for glycerine and also for liquid effluent is an area of research interest as well.

6.6.4.3 Fermentation and Treatment of Effluent

The bioconversion of glycerol into more valuable products, 1,3-propanediol, biobutanol (1-butanol) and succinic acid amongst others, has been reported in literature. Some of these routes have become commercial processes, currently employed in industry (Scheme 6.4).[118] From succinic acid, a wide range of

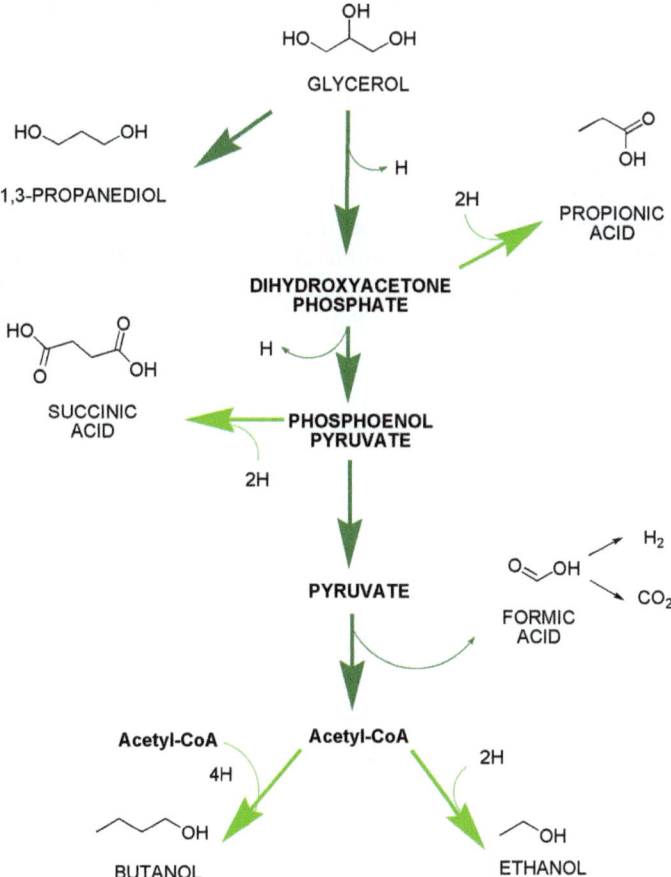

Scheme 6.4 Key biotransformation routes from glycerol.

chemical products including esters, amides, and alcohols can be obtained by using different chemical transformations.

Succinic acid has generally been produced from petrochemically derived feedstocks such as maleic anhydride. This type of production is highly polluting and there is increasing interest in the development of biobased commercial routes such as the one developed by the French company Bioamber,[119] which employs agricultural and food waste residues as a raw material for bioconversion of glucose into succinic acid. Other researchers are developing processes for the direct conversion of glycerol into this compound.[120]

In addition, biodiesel glycerine has been tested positively as a denitrifying agent in water and activated sludge treatment.[121] Nitrate, ammonia and nitrite are water pollutants, having the potential to promote eutrophication in water courses. Various types of bacteria (ammonia oxidising, nitrogen oxidising and heterotrophic, respectively) have the ability either to convert ammonia to nitrite/nitrate in aerobic conditions and into molecular nitrogen in anaerobic conditions, where nitrate acts as an electron acceptor (Scheme 6.5).[122] The process takes place in the presence of certain levels of organic matter (carbon source) as an electron donor. The biochemistry of the process is briefly explained in Scheme 6.6.[123]

Methanol has been commonly employed as a carbon source. Recently, glycerine has gained importance as a more economical and environmentally acceptable alternative to the toxic and flammable methanol.

Utilisation as a cement additive: Historically, the cement industry has employed a large number of additives (polyols, esters, amine and others) for different applications within the manufacture of the product (assisting grinding, improvement of flowability and giving beneficial effects on setting times).[124] Grace Ltd. and Professor Mario Pagliaro (University of Palermo, Italy)

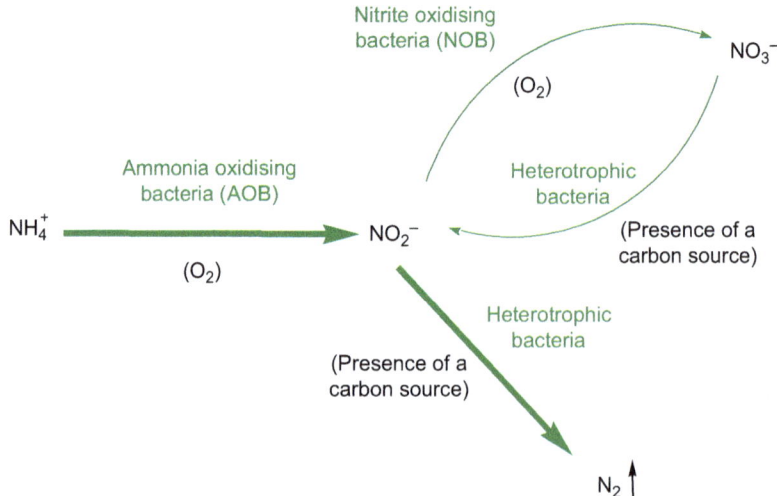

Scheme 6.5   General scheme for the denitrification chemistry.

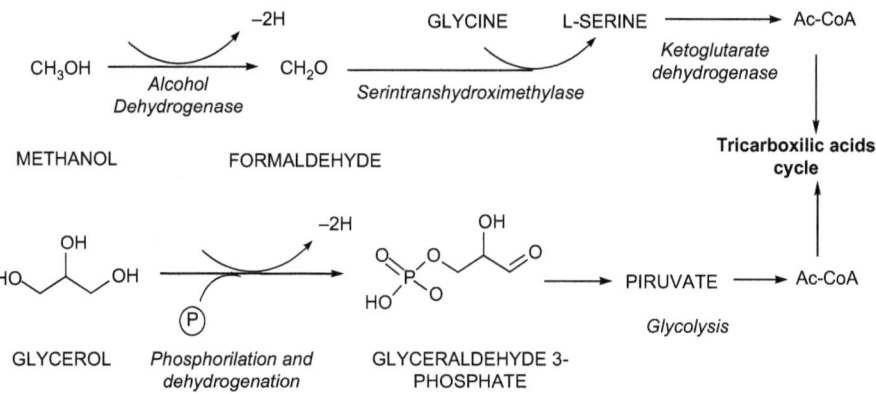

Scheme 6.6   Biochemical routes involved for the carbon source in denitrification processes.

recently patented the application of biodiesel glycerine in cement manufacture, where they established certain parameters for the composition, glycerol, MONG (material organic nonglycerol, *i.e.* organic compounds different from glycerol) and salt content.[125]

Production of alkyd resins: High-grade biodiesel glycerine has been employed in the production of alkyd aromatic resins. Glycerol is introduced into a phenol resin along with compounds such as polyvinylalcohol (PVA) using different acid catalysts. Not only does the use of glycerol improve the structural performance of the resin, but it also reduces cost. A purification process of biodiesel glycerine is usually required and in some of the examples for resin preparation, the residual methanol present in the parent glycerine is removed in the curing process.[126]

Solvent/Reaction media: recent studies have shown the potential for using biodiesel glycerine media in a number of chemical reactions. The preliminary research has shown increasing conversion and/or selectivity for transformations such as condensations, bioreductions or epoxide opening reactions. The authors of this work justify this fact by regarding the effects of some of the impurities present in crude glycerol as beneficial (surfactant activity of carboxylate species, solubility, *etc.*).[127]

6.6.5   Chemical Conversions of Glycerine

Some of the previously mentioned applications for glycerine allow biofuel producers to increase commercial margins on two fronts: by avoiding disposal and also by profiting from a biodiesel-coproduct. In the years 2000 and 2010 in the US National Renewable Energy Laboratory Report, glycerol was considered among the top platform chemicals for the production of added-value chemicals.[128] The structure of the compound and its versatility of chemistry indicate the potential of the compound to become the platform for a large number of industrial processes, depending on the availability of its supply, consistency and quality of the co-product. The successful production of chemicals from biodiesel glycerine would have an added economic and environmental value.

A number of relevant processes and transformations reported in literature (2005–2013) have been the objects of the current work (Figure 6.16). The study of these products and application has constituted the basis for suitable chemical strategies developed for the coproduct generated in biofuel production for a number of researchers.

In spite of the limitations of biodiesel glycerine as a feedstock for chemical conversions, much research has been conducted to establish suitable, selective catalytic routes towards glycerol products. Successful examples of glycerine in robust chemical processing have been developed, such as Solvay's process Epicerol® in the production of epichlorohydrin,[129] based on a chlorination process using hydrochloric acid the hydrogenolysis of crude glycerine to 1,2-PDO using Cu catalysts,[130] as well as the biosynthesis of 1,3-propanediol, butanol[118] and succinic acid, already mentioned. The following sections describe a number of chemistries, explored by Brocklesby Ltd. for biodiesel glycerine (Scheme 6.7).

Dehydration and oligomerisation of glycerol: one of the main features of the chemistry of glycerol is its decomposition at high temperature and acid conditions, forming various products, acrolein being one of the most relevant.

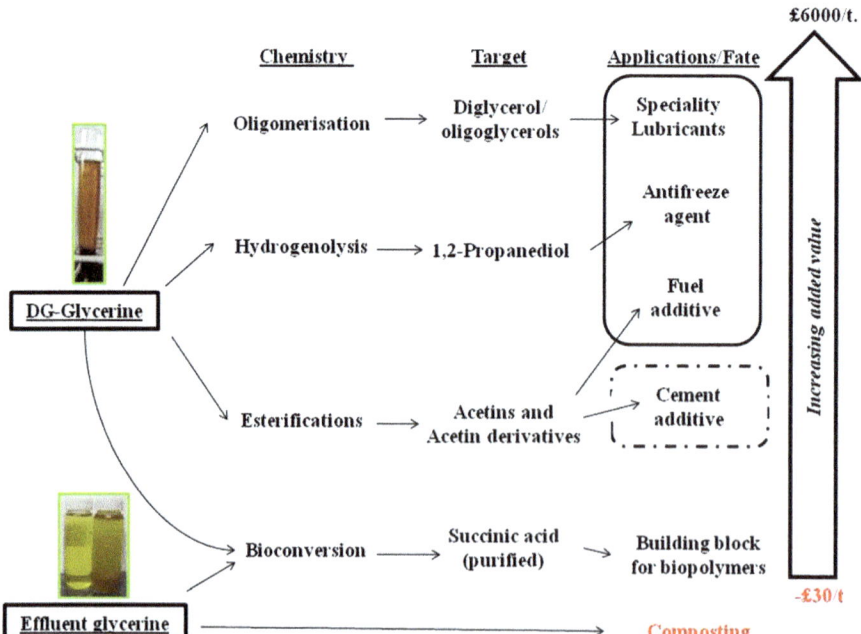

Figure 6.16   Strategy for glycerine at Brocklesby; chemistries, target products and applications.

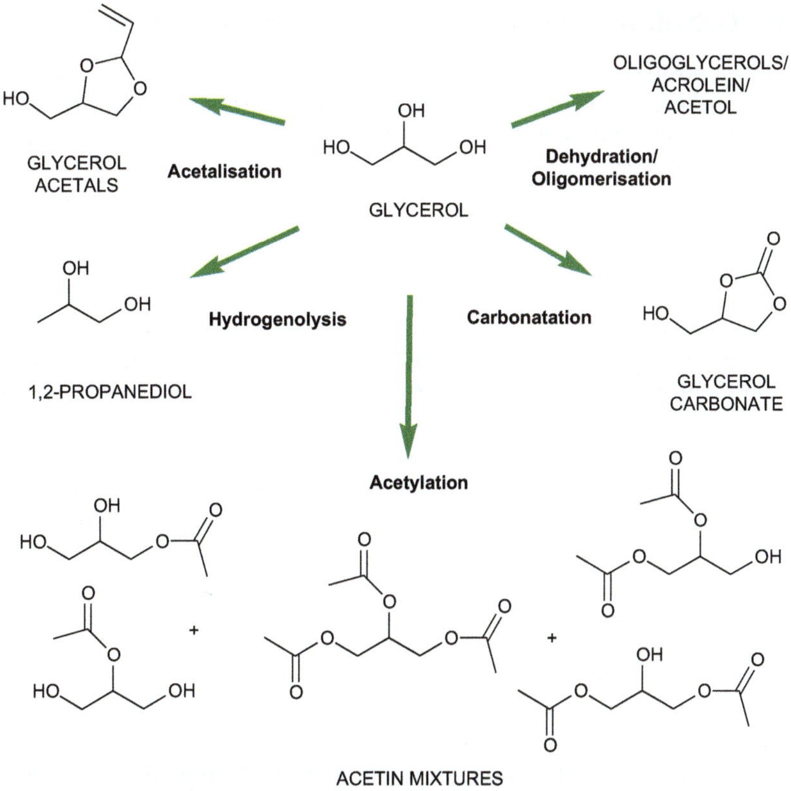

Scheme 6.7 Key chemistries explored.

Acrolein is a very interesting chemical from an industrial viewpoint where much investigation has been made to find a sustainable production route. The first parameters studied in glycerine were its stability at high temperatures and the product's thermal decomposition profile. Dehydration of glycerol can lead to compounds such as diglycerols and oligoglycerols, also at high temperatures. These chemicals can find applications as building blocks for polymers/specialty lubricants.[131]

Esterification/Acetylation; esters of glycerol have well-known applications within industry, for instance glyceryl acetates are well-known glycerol derivatives with known uses for instance, as fuel additives[132] or plasticisers.

Hydrogenolysis/Acetalisation: with focus on the production of 1,2-propanediol (1,2-PDO), and condensation products of glycerol with aldehydes. 1,2-PDO is employed in antifreeze/cooling formulations[133] and glycerol acetals are valuable chemicals with use as fuel additives.[134]

Carbonate synthesis; glycerol carbonate is a specialty chemical with applications as a solvent, epoxy resin and polymer precursor. The interest and markets for its production from crude glycerol have grown over the last few years.[135]

6.6.6 Dehydration of Glycerol to Acrolein and Acetol

The decomposition and internal dehydration of glycerol yield a number of products, where acrolein (2-propenal) and acetol (hydroxyacetone) are the most relevant chemicals from an industrial point of view. Acrolein has been employed in herbicide production over the last 60 years, with increasing demand. On the other hand, acetol is an intermediate for the synthesis of 1,2-PDO and the main dehydration by-product in the production of acrolein, primarily when using a heterogeneous acid catalyst. Acrolein, has traditionally been derived from propylene, however, the availability of lower-cost feedstocks such as biodiesel glycerine makes the development of renewable-based alternatives appealing.[136] The dehydration of glycerol has been reported to occur in the liquid and gas phase both with heterogeneous and homogeneous catalysts. The process is largely favoured by high temperatures (above 150 °C) and strong Brönsted acid conditions. In this respect, the use of a homogeneous catalyst in liquid phase gave the highest yield and selectivity to acrolein reported in the literature (to our knowledge), 84% when copper sulfate (in sulfolane) was employed[137] and 80% yield when using potassium hydrogen sulfate and potassium sulfate as a catalyst.[138] By contrast, a 50% yield was obtained under the best heterogeneous conditions employing phosphoric acid (5 wt%) supported on alumina.[139]

The development of full-scale processes based on this type of chemistry is still challenging, raising issues regarding a complex and discontinuous setup, in addition to safety concerns.

In the gas phase, recent examples in the literature report the use of heterogeneous catalysts with strong Brönsted acid sites, classified under increasing acidity measured with Hammet indicators (pKa (H_0) ≤ 6.8 the weakest acid up to pKa $(H_0) \leq -8.2$ for the strongest). Catalysts such as zeolites (pKa $(H_0) \leq -8.2$) and heteropolyacids $(-8.2 \leq$ pKa $(H_0) \leq -3.0)$ react with glycerol and water/glycerol feeds, giving excellent yields and selectivity to acrolein.[140] Heteropolyacids and heteropolysalts have given the highest acrolein yield for water/glycerol feeds. A recent work, using functionalised mesoporous materials with Cs heteropolyacids salts, studied the stability and the influence of the Brönsted/Lewis acidity in the acetol/acrolein and coke formation, linking acrolein and coke formation to strong Brönsted acidity.[141]

Corma *et al.* used highly acidic zeolites Y-zeolite doped with a ZSM-5 zeolite additive in a moving-bed reactor.[142] Acrolein is invariably obtained as the main product on changing temperatures and variable water content in the feed. At higher temperatures the selectivity is slightly shifted to the formation of acetaldehyde as one of the main by-products, nevertheless, acetol is still the main secondary product. The increase of glycerol concentration in the feed increases the production of coke by treatment at high temperature.

The extremely toxic nature of acrolein, however, is an issue to bear in mind in developing industrial production to the aldehyde and in order to minimise its formation as a by-product in other catalytic processes (esterifications, hydrogenations, *etc*).

6.6.6.1 Selectivity and Mechanism of Formation of Acrolein and Acetol

However important for process optimisation, research into catalyst selection should be coupled with a fundamental understanding of the reaction mechanism. Theoretical studies found 3-hydroxypropanal to be the most accessible product *via* (Brönsted) acid catalysed dehydration of glycerol ($E_a = 22.7$ kcal mol^{-1} compared to acetol $E_a = 24.9$ kcal mol^{-1}). 3-Hydroxy-propanal, is very unstable at high temperatures and undergoes a second dehydration to acrolein.[143] Furthermore, the investigation of acetol stability in the presence of catalysts with high numbers of Lewis sites demonstrated little formation of oligomers or coke under the same conditions that acrolein would do. In addition, the use of the same catalyst using glycerol as a feed demonstrated an increase of acetol formation with respect to the production of acrolein.[144] Therefore, there seems to be a correlation between the acid site type and the products generated, where the presence of Lewis acid sites in the catalyst will promote the dehydration of glycerol to acetol and the presence of Brönsted acid sites would favour acrolein formation, as shown in Schemes 6.8 and 6.9.[145]

The understanding of this mechanism might also be relevant to the production of other chemicals, where selectivity could be tuned to the production of glycerol (2,3-Epoxypropan-1-ol)[146] or allyl alcohol (2-propenol), for instance. The latter has been obtained with good selectivity and acceptable yields by using Fe$_2$O$_3$ nanoparticles.[147] The presence of hydroxyl groups on the surface of the catalyst and its topology seem to play a fundamental role in both the dehydration of glycerol to acrolein and subsequent hydrogen transfer with glycerol as a donor to the carbonyl functionality of acrolein reaching yields of 20–25%.

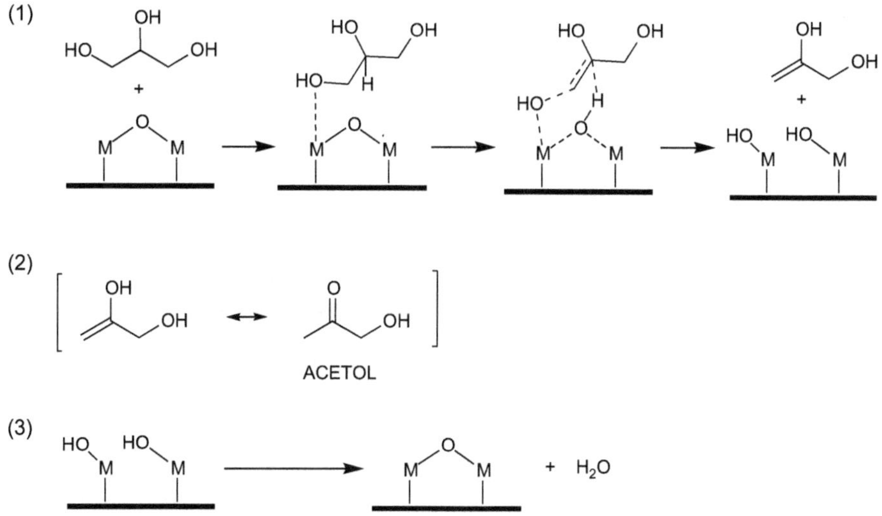

Scheme 6.8 Dehydration of glycerol in the presence of Lewis-acid catalyst.

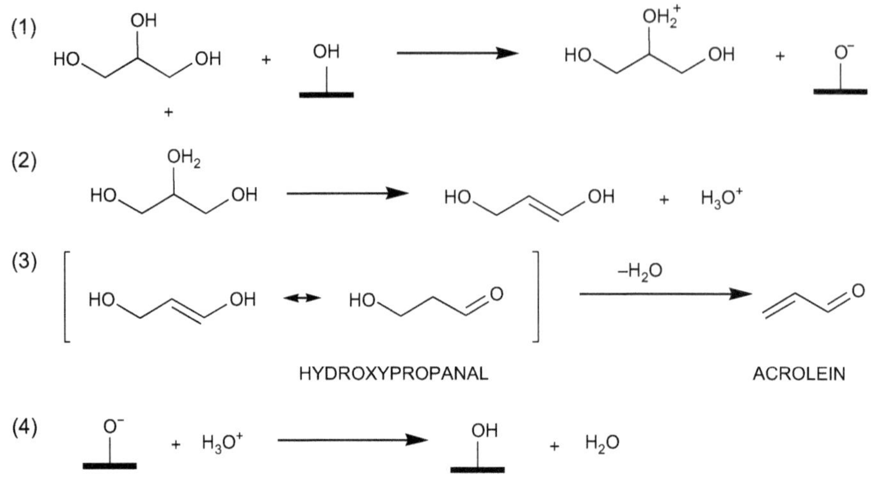

Scheme 6.9   Dehydration of glycerol with Brönsted-acid catalyst.

6.6.7   Oligoglycerol and Ethers

Oligomers, and alkyl ethers of glycerol might find applications in defined markets within food industries, as emulsifiers, additive stabilisers and fuel additives (Scheme 6.10).[148] Diglycerol and other oligoglycerols such as triglycerol and glycerol alkyl ethers are compounds with high commercial potential. There are a number of applications where they are employed as precursors of renewable-based lubricants, antifogging and antifoaming agents,[149] components of sunscreen formulations,[150] corrosion and cold-property improvers in mineral fuels and biofuels.[151]

Oligomerisation of glycerol, is promoted by strongly alkaline conditions and high temperatures (above 200 °C, generally 220–240 °C), using either homogeneous reagents with different solubility (Na_2CO_3, NaOH, CaO, MgO)[152,153] or functionalised silicas and mesoporous catalysts (Cs MCM-41).[154] The use of sodium or potassium carbonate is favoured over the equivalent hydroxides for the activation of glycerol. This is considered to be due to better solubility. The selectivity, however, is poor, yielding a complex reaction mixture leading to a wasteful work-up (Scheme 6.11). Nevertheless, examples of selective synthesis to linear oligomers have been shown with the use of high surface area activated CaO and MgO as catalysts with high conversion and selectivity to linear di- and triglycerol as shown in Scheme 6.13.[155]

Other examples relevant to this work are the production of methyl-glycerols, mono- and diethers. Methyl glycerols have reportedly been used as a cryopreservation agent[156] and as an intermediate for the synthesis of dioxolanes,[157] which are well-known fuel additives.[158] In addition the selective methylation of glycerol with dimethyl sulfate (DMST) in the presence NaOH has been studied by a number of authors.[159,160] Methyl-O-glycerols have found applications as diesel additives and as a cryogenation agent (Scheme 6.12).

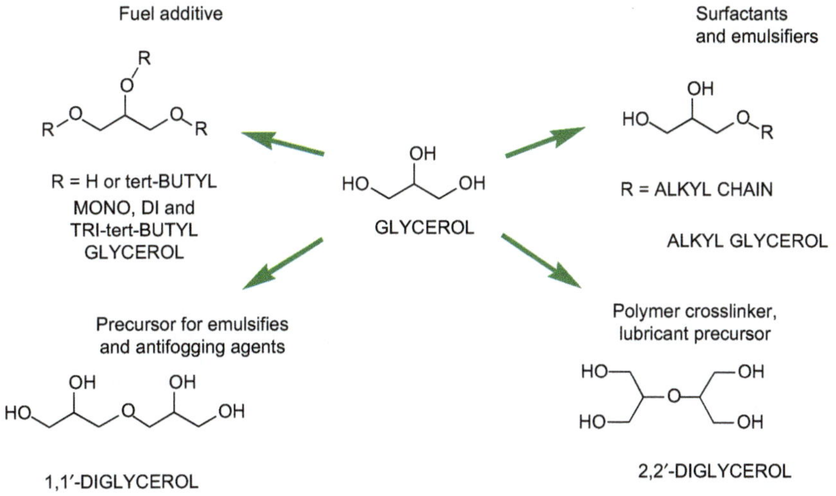

Scheme 6.10 Oligomerisations and etherifications of glycerol with Brönsted-acid catalyst.

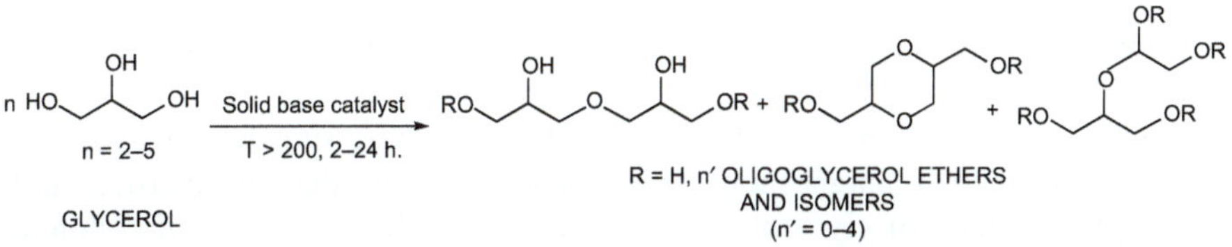

Scheme 6.11 General scheme for the oligomerisation of glycerol.

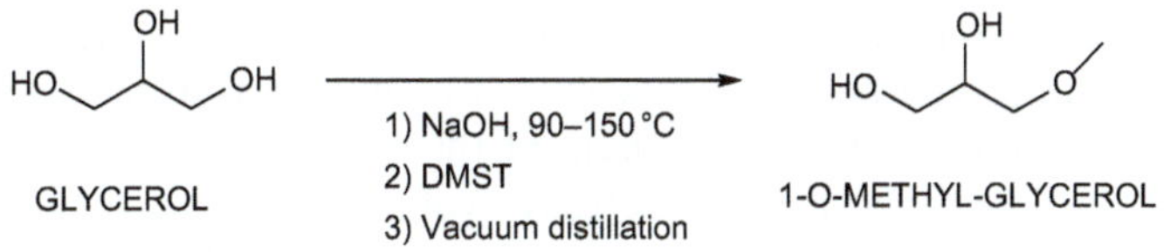

Scheme 6.12 Mechanism for the selective production of linear oligoglycerols.

Examples of the synthesis of ethyl glycerol with various solid acid catalysts[161] have been described, as well as other etherification processes using grafted silicas,[162] and zeolites.[163] The production of methyl glycerols from biodiesel glycerine seems interesting, due to the *in situ* ether generation of two major components of glycerine (glycerol and methanol).

6.6.7.1 Selectivity and Mechanism

CaO (and also MgO to a lesser extent) is a good example of a selective catalyst for the formation of linear oligoglycerols. The selectivity of the process might

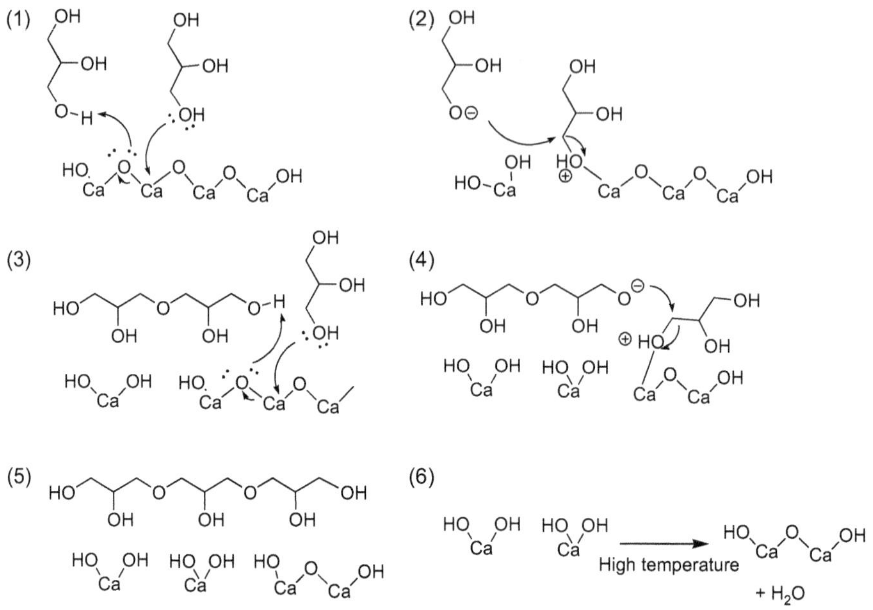

Scheme 6.13 Mechanism for the selective production of linear oligoglycerols.

be defined by a combination of available Brönsted base sites with Lewis acid sites in calcium (or magnesium) oxide promoting the dehydroxylation and nucleophilic attack of glycerol in these conditions; the formation of colloidal particles, by association with water generated *in situ*, seems to play an important role in the limitation of a further oligomerisation to branched or linear structures, as shown in Scheme 6.13.[164]

Selectivity and shape-dependent improvements for oligomerisation of glycerol have also been found in modified mesoporous silicas. Both the impregnation and incorporation of elements to mesoporous silicas presented problems of leaching and pore blocking of the catalyst, leading to low selectivity towards oligoglycerols (di- and triglycerol). By contrast, the exchange of the proton with caesium from the silanol groups present in MCM-41 and further calcinations of the resulting materials, increased the basicity of the catalyst, considerably improving the selectivity towards di- and triglycerol (Scheme 6.14) at high glycerol conversions.[165] This selectivity increase is related both to the fact that the catalyst is free from pore blocking, and also to the shape-limiting effect given by the mesopore, achieving more selective processes than those obtained with a homogeneous catalyst or with other solid catalysts such as zeolites.

6.6.8 Esterifications

A number of applications have been envisaged for the conversion of glycerine into another group of valuable products: glycerol esters and particularly

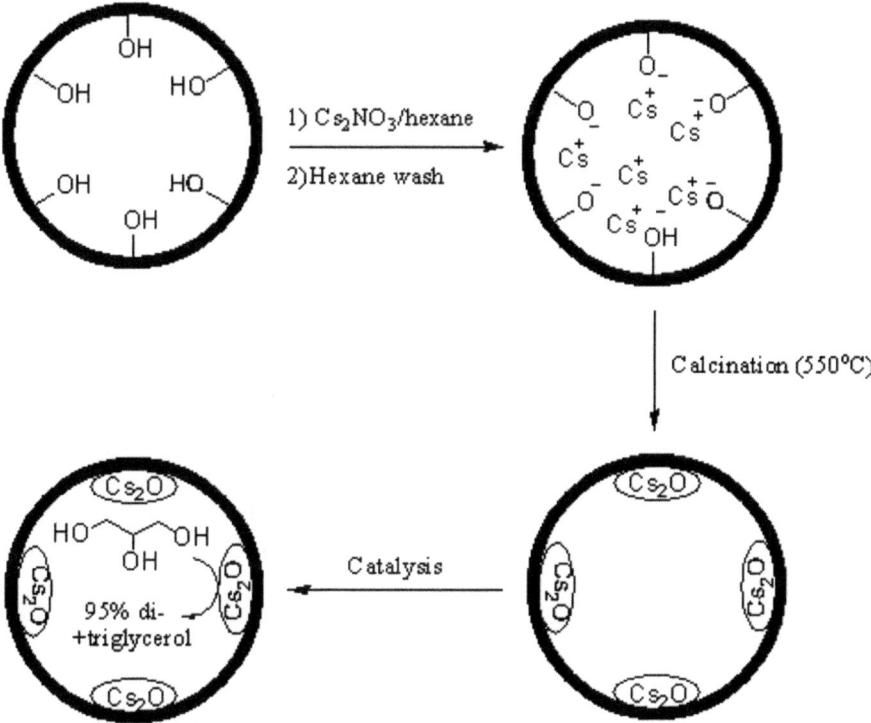

Scheme 6.14 Catalyst formation and selective oligomerisation of glycerol over modified silica.

glycerol acetates. Diacylglycerides (DAG) and monoacylglyceride (MAG), are widely used in the food industry,[166] and may find applications as additives in various other sectors.[167] Furthermore, compounds such as glycerol mono-oleate (GMO), can be derived either from glycerine or from used cooking oil. GMO is a commonly used engine antiwearing agent,[168] and its production from biodiesel glycerine would be relatively feasible either by UCO hydrolysis or esterification of glycerol. Another group of products with a high number of applications are the acetins or acetyl-glycerol esters,[169] which are interesting compounds with uses in fuels and biodiesel additives markets. Acetins increase performance of fuel (as an octane improver in gasoline), thus improving efficiency. A relevant example found in the patent literature is the transesterification of glycerol with ethyl and methyl acetates for the production of fatty acid (methyl/ethyl) esters. A range of proportions (0.5 to 10 wt%) of glyceryl triacetate (triacetin) is added to biodiesel with an improvement in the freezing-point properties.[170] Mixtures of acetins have also been employed as plasticity improvers/compression strength agents for the production of cement.[167] Using a nonpurified mixture has obvious potential for biodiesel glycerine derivatives compared with the use of a single refined product.[171]

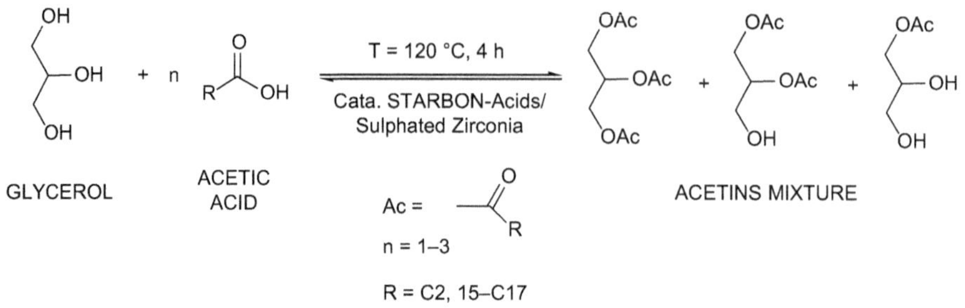

Scheme 6.15   General scheme for the acetylation of glycerine.

Acetylated mixtures (Scheme 6.15) require little or no purification. Mixed esters (acetic acid/long-chain free fatty acids) are also relevant because they may have improved properties from the parent glycerine in terms of hydrophobicity, or plasticizing ability.[172] Our research interest has focused on the use of heterogeneous catalysts for the direct production of acetins with either acetic acid or acetic anhydride.

6.6.8.1   Selectivity and Mechanism

A number of heterogeneous catalysts have been used in esterifications of glycerol with different lengths of fatty acids, with alkaline oxides and hydroxides (MgO, Ca(OH)$_2$, NaOH),[173] mineral (sulfuric acid), zeolites, heteropolyacids, zirconias, organomodified silicas and carbonaceous materials as main examples of catalysts. The influences on conversion and selectivity are variable; the most relevant, however, involves the relative affinity of the substrate to the catalysts and the influence of temperature and high molar ratio acid/glycerol and temperature on the extent of reaction and increasing degree of substitution. This is shown in the acetylation of glycerol with hybrid organosulfonic acid silicas and mesoporous carbonaceous materials through microwave activation.[174] In the first paper, selectivity patterns were described, successfully developing a multidimensional model to optimise conditions for the production of mixtures of acetylated glycerol.

There are also good examples of selective acetylation of glycerol in the literature, using heteropolyacids.[175] Other examples of selective processes to triacetin have been reported either using acetic anhydride, or in a combined esterification/acetylation process.[176] The mechanisms for the esterification have been found to vary according to conditions. Scheme 6.16 shows the mechanism proposed by Jacobs and coworkers, when modified silicas are employed to selectively produce monoglycerides.[177]

Similar shape-dependent processes have been reported by Corma *et al.*,[178] Jérôme *et al.*[179] and others[180] by using hybrid organo-modified silicas and hydrotalcites amongst other heterogeneous catalysts, and bare fatty acids or methyl esters as acylating agents. For these types of catalyst and conditions, the

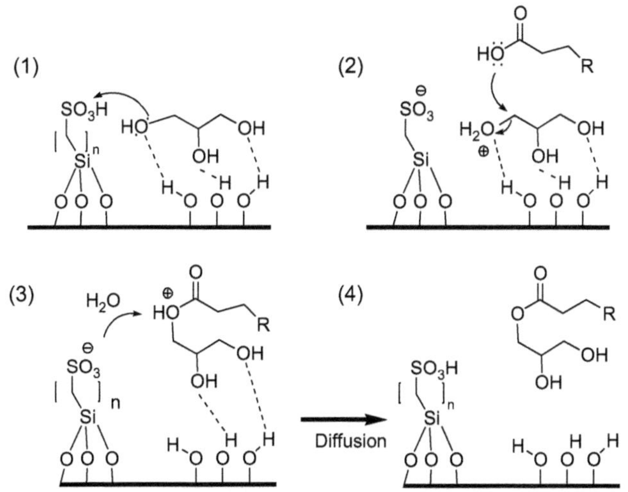

Scheme 6.16 Mechanism of acylation of glycerol over functionalised silicas.

mechanism described is highly surface dependent. Silicas without functionalisation have a strong affinity for glycerol, however, the introduction of organosulfonate acid groups dramatically changes the adsorption/desorption and diffusion dynamics of reagents and products. The surface nature of the catalyst allows the single reaction of a moiety from glycerol (Scheme 6.16), with the selectivities increasing as the chain length of the FFA increases to 12 to 18C. When the product is formed, the affinity of the compound for the catalyst changes, diffusing more easily from the active site, leaving it free for an additional glycerol moiety to react. The selectivity would be expected to decrease in the same conditions if acetic acid is employed instead of the free fatty acids, due to the reduction of lipophilicity of the acid.

Barrault also justifies the introduction of ionic liquids in the silica matrix to increase the stability and catalyst life, which is especially relevant when using biodiesel glycerine as the feed for the acetylation process.[181]

6.6.9 Hydrogenolysis

Hydrogenolysis of glycerol is selectively catalysed by heterogeneous catalysts such as $CuCrO_3$ and Pd/C system. Others systems described in literature lead to variable yields and selectivity (Pt/C, Cu/alumina, Raney Ni, Raney Cu, Ru/alumina, *etc.*).[182] The mechanism proposed by Suppes and coworkers comprises the conversion of glycerol to acetol as a first step and the hydrogenation of the ketonic compound.[183] Acetol formation in contrast, is generally favoured by the presence of Lewis sites (Scheme 6.7), so it is produced as the main product under low hydrogen pressure and in the presence of catalysts such as $CuCrO_3$, Pd/C, CuO-ZnO at temperatures above 180 °C.[182] This mechanism would rule out previously proposed reaction paths such as the formation of glyceraldehyde as an intermediate.[184]

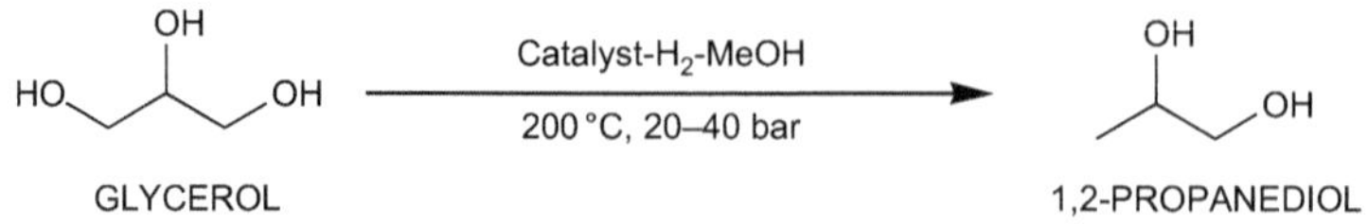

Scheme 6.17   General conditions for the hydrogenolysis of glycerol.

1,2-PDO has been obtained from crude glycerol using copper chromite and copper/carbon as catalysts, with good results (Scheme 6.17). The use of $CuCrO_3$ as the preferred catalyst is justified by the high selectivities and conversions reported by Suppes and others. The process involved a two-step reaction, wherein glycerol is converted first to pure acetol, at 0.65 bar and consequently into glycol by changing the pressure (up to 13.8 bar) in the presence of the same catalyst. The reaction has proved to be a truly reactive distillation process. The use of a single-step process proves to be difficult in achieving selectivities higher than 80% with the surface of the catalyst becoming coated by carbon deposits/oligomers.

$CuCrO_3$ has proven to be a suitable catalyst for the hydrogenation of biodiesel glycerin into 1,2-PDO, however, the removal of salts (chlorides, sulfates and/or phosphates) is likely to be required in order to avoid $CuCrO_3$ catalyst poisoning as much as possible.[185]

The isomer 1,3-propanediol is also of interest, with potential in the fibre and polymer industries; it is produced primarily by biochemical transformation from crude glycerol.[186] The market for these glycols is expected to grow over the next few years, with an estimated annual global demand of 535 000–717 000 tonnes and growth of 4%/pa.[187]

Davy Ltd. and Eastman Chemicals Co. have filed at least two patents on the chemical hydrogenation of crude glycerol for the production of 1,2-propanediol.[188] In these patents, the terminology "glycerol from any source" implies the use of biodiesel glycerine for this invention. There is no mention of a pretreatment for the feedstock in these examples. By contrast, biodiesel glycerine is clearly acknowledged in a patent where it is claimed that water/methanol solutions of minimum 40 wt% content of glycerol[189] are reacted in the presence of mixed oxide ($CuO–CeO_2–SiO_2$) at high temperatures and pressure.

6.6.10   Acetalisation and Hydrogenolysis

Hemicetals and acetals of glycerol have a number of interesting properties, from fuel additives to solvents, synthons for pharmaceutical production and other uses. Different approaches to these products have been taken in the past, with the use of homogeneous catalysts such as p-toluenesulfonic acid as a traditional route to deriving cyclic acetals from glycerol and other carbohydrates.[190]

The possibility of *in situ* formation of carbonyl compounds from glycerine and their sequential acetalisation with glycerol is an interesting approach. This

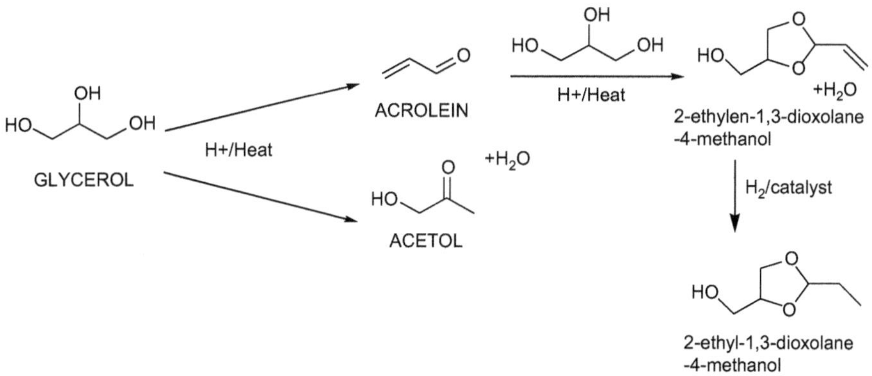

Scheme 6.18 General conditions for the one-pot 1) dehydration/ acetalisation; 2) hydrogenolysis of glycerol.

would be achieved by reacting the triol into acrolein or acetol at high working temperatures followed by further condensation with the remaining glycerol, as shown in Scheme 6.18.

The synthesis would combine the dehydration of glycerol to a carbonyl compound with the formation of acetal/hydrogenation or *vice versa*.

Both the hydrogenation/acetalisation and condensation of glycerol with carbonyl compounds have been reported in literature. Pd/C catalysts have been employed for the hydrogenolysis of glycerol and conversion into hemiacetals at mild temperature and H_2 pressure conditions.[191] Another interesting process published by Mota is the use of zeolites with medium Si/Al ratio leading to the acetalisation of glycerol with carbonyl compounds such as acetone or formic acid.[192] In addition, Peinado also reported a combined acetalisation/ esterification process using acid conditions and producing interesting chemicals for potential use as fuel additives.[193]

6.6.11   Glycerol Carbonate

Glycerol carbonate is another chemical worthy of note. It has the potential to substitute petrochemically derived propylene carbonate in applications such as a green solvents, or precursors for epoxy resins and polyurethanes, (polyurethane market demand is estimated to be £4.9 billion in the U.S. alone[194] and it keeps increasing in emerging economies). Glycerol carbonate can also be employed for the production of oligoglycerols and oligoglycerol esters (with application as antifogging agents) and for the production of chemicals such as glycidol. Its current production involves the hydrolysis of epichlorohydrin, *via* the glycerol (Epicerol process), or from propylene. Established routes to glycerol carbonate from glycerol may involve the use of noxious chemicals such as phosgene,[195] therefore new and greener routes are desirable. Examples of such routes could include the use of organic carbonates [Dimethyl carbonate (DMC): (Scheme 6.19)], supercritical CO_2 (Scheme 6.21) or urea (Scheme 6.22)

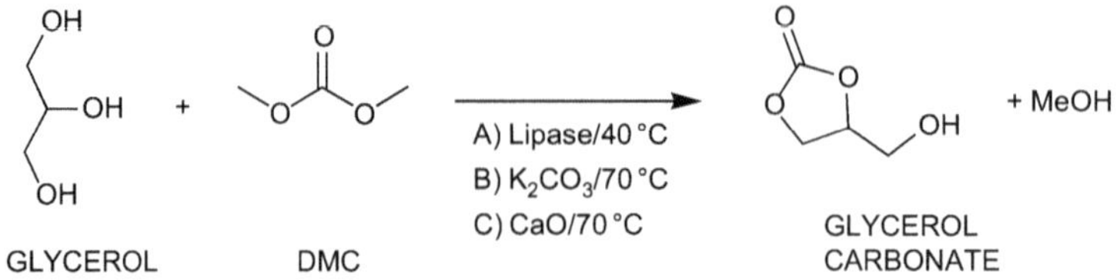

Scheme 6.19 DMC in the synthesis of glycerol carbonate.

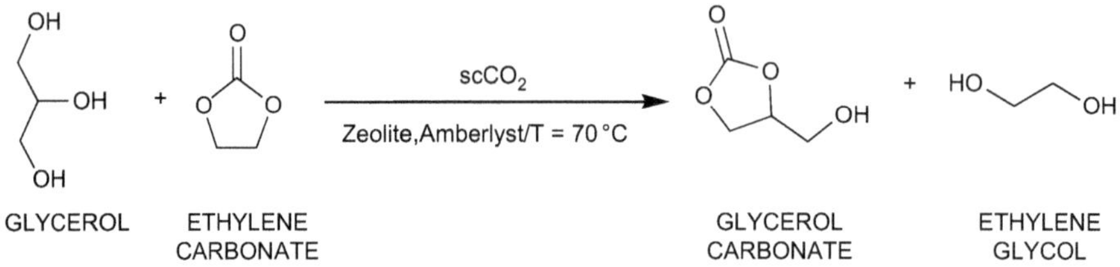

Scheme 6.20 Glycerol carbonate synthesis in supercritical CO_2.

as carbonating agents catalysed by solid bases such as metal oxides, carbonates, Zn-derived Lewis acids ion-exchange resins, modified zeolites or biocatalysts such as lipases.

The use of organic carbonates is also appealing, since the production of the carbonate could generate valuable byproducts such as methanol, which can be recovered and recycled back into biodiesel production. An example that illustrates the use of $scCO_2$ and organic carbonate as a reagent is the trans-esterification of glycol carbonate with glycerol, occurring in the presence of ion exchange resins and zeolites (Scheme 6.20). This work is of value from an environmental perspective but the extent of the reaction and the molar excess of reagents render it unviable from a commercial point of view.

The carbonation of glycerol with CO_2 and CO/H_2 is highly attractive from an environmental stance. Researchers have described successful examples of conversion to glycerol carbonate by using Sn catalyst (Scheme 6.21).

Glycerol carbonate is much more reactive than glycerol, primarily towards nucleophilic additions, having four different nucleophilic centres (one of them pro-chiral). This factor, despite being advantageous in other synthetic routes, is the main drawback for its synthesis. Previous research has shown the occurrence of substitution and elimination reactions depends on the conditions employed and is important to bear this in mind when developing a synthesis with these products.[196]

6.6.12 Other Transformations

Oxidations are relevant because they can lead to a wide number of monomers that can be used as substitutes for oil-derived chemicals, however, the markets for such products, with the exception of DHA (dihydroxy acetone), are not

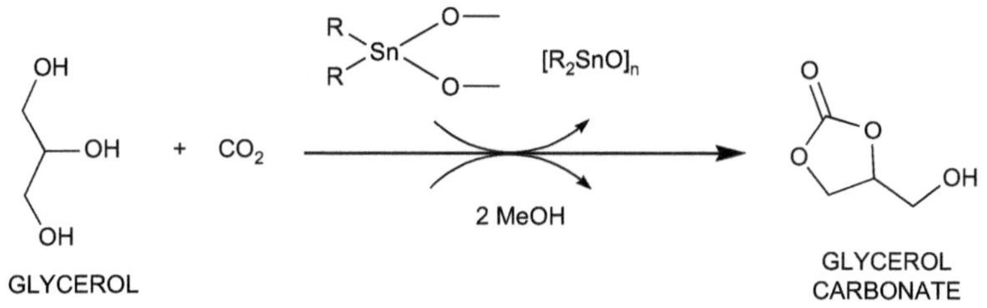

Scheme 6.21 Glycerol carbonate synthesis with CO_2 and tin catalyst.

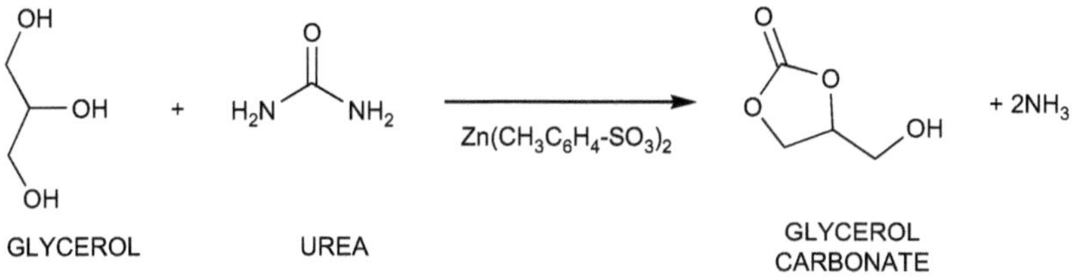

Scheme 6.22 Glycerol carbonate synthesis from urea in the presence of a zinc catalyst.

fully developed. Selective catalytic conversions of glycerol, therefore, are promising areas of research with great potential for the development of chemicals with new applications.[197] These transformations have not been covered in this project (Scheme 6.22).

6.7 Conclusions

The current work has explored the potential for valorising reclaimed vegetable oils/fats and similar food related co-products into multiple applications.

As shown, most of these materials generated in the fats and vegetable oil production supply chain are recycled as animal feed supplement (ca. 90%, more than 120 m tonnes worldwide). However up to 20 m tonnes of these by-products are unsuitable for feed (mainly catering waste, and UCO), and have the potential to be employed in technical applications. This group of by-products tend to be used in biofuel production, with biodiesel, biogas and, to a lesser extent, hydrocracking to being their main end-uses. However there is potential to gain more value by developing other products from these feedstocks, which explains the enormous amount of research currently being carried out in this area to:

- Find alternative feedstocks for biofuels; such as rendered fat from miscellaneous waste, sludges, non edible tallow and fats, *etc.*
- Achieve more efficient and greener biofuel processes using enzymes and heterogeneous catalysts.
- Valorise fats and oil co-products such as soap-stocks and biodiesel glycerine into fuels and chemicals.

Indeed, alternative routes which generate potentially higher added value products than fuel have been developed over the years for oils and fats co-products, including lubricants, varnishes and the production of surfactants. However, despite the amount of work conducted in this area, the utilisation of the feedstocks by the chemical industry (other than biofuel production) remains challenging. Their specification and stability is generally variable and therefore considered unreliable by chemical producers. Also, despite the development of the "end-of-waste" regulations, there are considerable administrative constraints and uncertainties for producers in employing these materials in the manufacturing industry. Catering co-products are generally classed as waste and therefore their use must safeguard human and environmental health as dictated by the relevant environmental regulations and REACH. In this respect the production of bio fuel and biogas has a clear advantage with regard to their acquisition of "end-of-waste" status.

Nevertheless, the various stakeholders of the oils and fats supply chain; manufacturing, consumers, researchers and regulators need to continue co-operating in order to develop more sustainable and resource efficient supply chains, from seed crushing, fat rendering to food production, ideally generating added value products and fuels at the end-of-life stage for the co-products.

How about a high performance UCO-derived HOSO bio-lubricant for biodiesel-fuelled tractors harvesting these crops?

References

1. (a) C. S. K. Lin, L. A. Pfaltzgraff, L. Herrero-Davila, E. B. Mubofu, S. Abderrahim, J. H. Clark, A. Koutinas, N. Kopsahelis, K. Stamatelatou, Fi. Dickson, S. Thankappan, Z. Mohamed, R. Brocklesby and R. Luque, *Energy Environ. Sci.*, 2013, **6**(2), 426–464; (b) M. R. Kosseva, Processing of food wastes: Chapter 3, *Adv. Food Nutr. Res.*, 2009, **58**, 57–136.
2. Proteins from oilseeds – Nick Bajjalieh, Ph.D., Available online at: http://www.fao.org/docrep/007/y5019e/y5019e09.htm [Accessed on 03.02.13].
3. F. Rosillo-Calle, L. Pelkmans and A. Walter, A Global Overview of Vegetable Oils, with Reference to Biodiesel, *A Report for the IEA Bioenergy Task 40, Imperial College London (UK), VITO (Belgium) and UNICAMP (Brazil), June 2009*.
4. (a) F. D. Gunstone, in *Oleochemical Manufacture and Applications*, edited by F. D. Gunstone and R. J. Hamilton, Taylor & Francis, Sheffield, 2001, Chapter 1, pp. 1–22; (b) K. Hill and R. Höfer in *Sustainable Solutions for Modern Economies*, edited by R. Höfer, RSC Publishing, Cambridge, 1st edn., 2009, ch. 9.1, pp. 167–228.
5. BP Statistical Review of World Energy, June 2012 [Online]. Available: http://www.bp.com/assets/bp_internet/globalbp/globalbp_uk_english/reports_and_publications/statistical_energy_review_2011/STAGING/local_assets/pdf/statistical_review_of_world_energy_full_report_2012.pdf [Accessed: 05.10.12].

6. *Feeding Fats Safety*, Report For Fat/Feed Producers Deliverable no. 12 Paolo Bondioli, Deliverable no. 12. "Feeding Fats Safety" and registered with the EU code FOOD-CT-2004-007020.

7. Landfill directive Regulatory Guidance Note 14 (version 2.4) on The Duty of Care and the European Waste Catalogue, UK Environment Agency, 20 August 2007.

8. Scottish Environment Protection Agency, National Best Practice Project Phase 1: *Used Cooking Oils*, 6 September 2005.

9. Directive 2006/12/EC of 5 April 2006 on waste.

10. *Household Food and Drink Waste in the UK*, WRAP, UK, November 2009 [Online]. Available: http://www.wrap.org.uk/sites/files/wrap/Household%20food%20and%20drink%20waste%20in%20the%20UK%20-%20report.pdf [Accessed: 05.10.12].

11. G. Mangesh, G. Kulkarni and A. K. Dalai, *Ind. Eng. Chem. Res.*, 2006, **45**(9), 2901–2913.

12. D. Stevens, *Waste Vegetable Oil Recycling for Bio-diesel Production in Essex & Cambridgeshire*, Waste WISE Overview Report, 2 May 2003.

13. D. G. Tiffany, *The Growth of Alternative Fuels: Minnesota and U.S. Perspectives*, Submitted for a Joint Conference of the University of Minnesota, University of Padova and University of Bologna, Wisconsin, 2002.

14. W. Jinfu, in Better Engines than Table, *China Daily*, 31.10.2011 [Online]. Available: http://www.chinadaily.com.cn/opinion/2011-10/31/content_14004907.htm [Accessed on 01.09.12]; (b) D. Barboza, in Recycled Cooking Oil Found to Be Latest Hazard in China, The New York Times, 175, 31.03.2010 [Online]. Available: http://www.nytimes.com/2010/04/01/world/asia/01shanghai.html?_r¼1 [Accessed: 01.09.12].

15. PDM Group, *Tallow and rendered animal fat*, News Report, 2008, [Online]. Available: http://www.pdm-group.co.uk/news/2008/tallow_rendered_animal_fats.html [Accessed: 11.07.12].

16. Available: http://www.fda.gov, (a) Guidance for Industry: Action Levels for Poisonous or Deleterious Substances in Human Food and Animal Feed and; (b) Guidance for Industry: Small Entities Compliance Guide for Renderers-Substances prohibited for Use in Animal Food or Feed. [Accessed: 23.06.13].

17. ADAS UK Ltd., *West Midlands Non-Food Crops Opportunities/Mapping Study*, 15th October 2007.

18. Available under subscription: F.O. Licht's World Ethanol and Biofuels Report, weekly on-line: http://www.agra-net.com/portal2/showservice.jsp?servicename¼as072# [Accessed: 12.08.12].

19. R. G. Cooper and J. M. Rusell, in The *New Zealand meat processing industry, present effluent treatment practice and future directions*, Proceedings of the 1992 Food Industry and Environmental Conference, The Georgia Tech Research Institute, Atlanta, p. 95.

20. L. C. Meher, D. Vidya Sagar and S. N. Naik, *Renew. Sust. Energ. Rev.*, 2006, **10**, 248–268.

21. Available online at: http://www.clearfleau.com/index?id¼18. [Accessed: 12.08/2012].
22. M. R. Kosseva, C. A. Kent and D. R. Loyd, *Biochem. Eng. J.*, 2003, **15**, 125–130.
23. (a) M. J. Haas, *Fuel Process Tech.*, 2005, **86**, 1087–1096; (b) A. L. Tiwari, A. Kumar and H. Raheman, *Biomass Bioenerg.*, 2007, **31**, 569–575.
24. S. P. Kochhar and C. Gertz, *Eur. J. Lipid Sci. Technol.*, 2004, **106**, 722–727.
25. Y. Catel, F. Aladedunye and R. Przybylski, *JAOC*, 2012, **89**, 55–66.
26. J. B. Rosseli, *Grasas y Aceites*, 1998, **49**(3–4), 282–295.
27. S. Paul and G. S. Mittal, *Crit. Rev. Food Sci. Nutr.*, 1997, **37**(7), 635–662.
28. S. Hara, E. Ogawa and Y. Totani, *J. Oleo. Sci*, 2006, **55**(4), 167–172.
29. S. Marmesat, E. Rodrigues, J. Velasco and C. Dobarganes, *Int. J. Food Sci. Technol.*, 2007, **42**, 601–608.
30. "*Standard Test Method for Water Using Volumetric Karl Fischer Titration*", ASTM E203–08.
31. "*Animal and vegetable fats and oils - Determination of acid value and acidity*". ISO 660:2009.
32. "*Standard Test Method for Acid Number of Petroleum Products by Potentiometric Titration*"- ASTM D664–11a.
33. Rojas, Narvaez, *Ingeniería e Investigación*, vol. 31, No. 1, April 2011 (pp. 83–92).
34. "*Fat and oil derivatives-fatty acid methyl esters (FAME); Determination of ester and linolenic acid methyl ester contents*", EN14103:2003, 2003.
35. http://lipidlibrary.aocs.org/frying/a-polar/index.htm.
36. "*Fat and oil derivatives. Fatty acid methyl esters (FAME). Determination of oxidation stability (accelerated oxidation test)*", BS14112:2003, 2003.
37. C. Ceballos and H. Fernández, *Food Res. Int.*, 2000, **33**, 357–365.
38. E. Coni, E. Podestá and T. Catone, *Thermochimica Acta*, 2004, **418**, 11–15.
39. Afaf Kamal-Eldin, *Eur. J. Lipid Sci. Technol.*, 2006, **58**, 1051–1061.
40. In a Communication of 18 May 2010 on future steps in bio-waste management in Europe.
41. *Pioneering development set to increase food recycling in the UK*, Association for Organics Recycling, News 2012 [Online]. Available: http://www.organics-recycling.org.uk/page.php?article¼2379&name¼ Pioneeringþdevelopmentþsetþtoþincreaseþfoodþwasteþrecycling þinþtheþUK [Accessed: 29.08.12].
42. Regulation (EC) No. 1907/2006 (as amended) of the European Parliament and the Council of 18 December 2006 concerning the Registration, Evaluation, Authorisation and Restriction of Chemicals (REACH).
43. (a) China - Order No. 7 "Provisions on the Environmental Administration of New Chemical Substances" 15 October 2010 (known as "China REACH") and Decree 591, "Regulations on the Safe Management of Hazardous Chemicals in China" (2011), effective 1 December 2011.

44. A. Rojas-González, O. Chaparro-Anaya, C. Andrés-Ospina, Assessment of Biodiesel-Diesel Blends for Producing Electric Energy, Ing. Univ. Bogota' (Colombia), **15**(2): 319–336, julio–diciembre de 2011. ISSN 0123-2126.

45. J. Janaun and N. Ellis, *Renew. Sust. Energ. Rev*, 2010, **14**, 1312–1320.

46. M. Cetinkaya and F. Karaosmanoglu, *Energ. Fuel*, 2004, **18**, 1888–1895.

47. R. Luque and J. H. Clark, *ChemCatChem*, 2011, **3**, 594–597.

48. Q. Shu, Z. Nawaz, J. X. Gao, Y. H. Liao, Q. Zhang, D. Z. Wang and J. F. Wang, *Bioresour. Technol.*, 2010, **101**, 5374–5384.

49. M. Kouzua and J. Hidaka, *Fuel*, 2012, **93**, 1–12.

50. (a) R. Jothiramalingam and M. Kuang Wang, *Ind. Eng. Chem. Res.*, 2009, **48**, 6162–6172; (b) A. P. Singh Chouhan and A. K. Sarma, *Renew. Sust. Energ. Rev.*, 2011, **15**, 4378–4399.

51. N. Viriya-empikul, P. Krasae, W. Nualpaeng, B. Yoosuk and K. Faungnawakij, *Fuel*, 2012, **92**, 39–244.

52. B. Zhang, Y. Weng, H. Xu and Z. Mao, *Appl. Microbiol. Biotechnol.*, 2012, **93**, 61–70.

53. S. Kent Hoekman, A. Broch, C. Robbins, E. Ceniceros and M. Natarajan, *Renew. Sust. Energ. Rev.*, 2012, **16**, 143–169.

54. M. V. Ruiz-Méndez, S. Marmesat, A. Liotta and M. C. Dobarganes, *Grasas y Aceites*, 2008, **59**, 45–50.

55. *Rancidity of Foods*, edn. J. C. Allen and R. J. Hamilton, 2nd edn., 1989, Elsevier Applied Sciences.

56. RFA Report on Year Two of the RTFO, and verified Data from 2009/2010, Renewable Fuels Agency, UK Goverment, 2010.

57. N. Zyaykina, V. Van Hoed, W. De Greyt and R. Verhé Zyaykina, *Lipid Technol.*, 2009, **21**, 182–185.

58. (a) K. Murata, Y. Liu, M. Inaba and I. Takahara, *Energy Fuels*, 2010, **24**, 707–717; (b) G. W. Huber, P. O'Connor and A. Corma, *Appl. Catal. A*, 2007, **329**, 120–129.

59. W. Charusiri, W. Yongchareon and T. Vitidsant, *Korean J. Chem. Eng.*, 2006, **23**, 349–355.

60. W. Charusiri and T. Vitidsant, *J. Energy*, 2003, **5**, 58–68.

61. W. Charusiri and T. Vitidsant, *Energ. Fuels*, 2005, **19**, 1783–1789.

62. US Patent, 7 232 935, 2007.

63. H. Aatola, M. Larmi, T. Sarjovaara and S. Mikkonen, Hydrotreated Vegetable Oil (HVO) as a Renewable Diesel Fuel: Trade-off between NOx, Particulate Emission, and Fuel Consumption of a Heavy Duty Engine, 2008 [Online]. Available: http://www.biofuelstp.eu/downloads/ SAE_Study_Hydrotreated_Vegetable_Oil_HVO_as_a_Renewable_Diesel_ Fuel.pdf [Accessed 03.12.12].

64. K. Sunde, A. Brekke and B. Solberg, *Energies*, 2011, **4**, 845–877.

65. (a) S. Berzegianni, A. Dimitriadis, A. Kalogianni and K. G. Knudsen, *Ind. Eng. Chem. Res.*, 2011, **50**, 3874–3879; (b) S. Berzegianni, A. Dimitriadis, A. Kalogianni and P. Pilavachi, *Bioresour. Technol.*, 2010, **101**, 6651–6656; (c) S. Berzegianni, A. Dimitriadis and A. Kalogianni, *Bioresour. Technol.*, 2010, **101**, 7658–7660.

66. (a) DIRECTIVE 2008/98/EC of the European Parliament and of the council of the 19 November 2008 on waste and repealing certain Directives; (b) UK Enviroment Agency, WRAP, March 2011.
67. S. Woodgate and J. van der Veen, *Biotechnol. Agron. Soc. Environ.*, 2004, **8**(4), 283–294.
68. V. L. Budarin, J. H. Clark, B. Lanigan, P. Shuttleworth, S. W. Breeden, A. J. Wilson, D. J. Macquarrie, K. Milkowski, J. Jones, T. Bridgeman and A. Ross, *Biores. Technol.*, 2009, **100**, 6064.
69. J. Petran, L. Pedišić, M. Orlović, Š. Podolski and V. Bradač, *GOMABN*, 2008, **47**(6), 463–478.
70. (a) S. Boyde, *Green Chem.*, 2002, **4**, 293–307; (b) R. Garcés, E. Martínez-Force and J. J. Salas, *Grasas y Aceites*, 2011, **62**, 21–28.
71. J. Bentley, in *Oleochemical Manufacture and Applications*, eds. F. D. Gunstone and R. J. Hamilton, Sheffield Academic Press, 1st edn., 2001, ch. 6, pp. 164–193.
72. C. Herrmann, J. Hesselbach, R. Bock, A. Zein, G. Öhlschläger and T. Dettmer, *CLEAN-Soil Air Water*, 2007, **35**, 427–432.
73. F. D. Gunstone, *Lipid Technol.*, 2012, **24**(4), 96.
74. P. Nagendramma and S. Kaul, *Renew. Sustain. Energy Rev.*, 2012, **16**, 764–774.
75. Y. M. Shashidhara and S. R. Jayaram, *Tribol. Int.*, 2010, **43**, 1073–1081.
76. G. Mendoza, A. Igartua, B. Fernandez-Diaz, F. Urquiola, S. Vivanco and R. Arguizoniz, *Grasas y Aceites*, 2011, **62**, 29–38.
77. A. A. Hayder, M. Y. Rosli, H. N. Abdurrahman and M. K. Nizam, *Int. J. Phys. Sci.*, 2011, **6**, 4695–4699.
78. F. D. Gunstone, J. Alander, S. Z. Erham, and B. K. Sharama, in *The Lipid Handbook with CD-ROM*, edited by D. Gunstone, J. L. Harwood, A. J. Dijkstra, CRC Press, 3rd edn, 2007, ch. 9, pp. 591–635.
79. A. Bono, O. P. Pin and C. P. Jiun, *J. Appl. Sci.*, 2010, **10**, 2508–2515.
80. S. Arumugam and G. Sriram, *Tribol. Trans.*, 2012, **55**, 438–445.
81. (a) V. Dossat, D. Combes and A. Marty, *J. Biotechnol.*, 2002, **97**, 117–124; (b) R. N. M. Kamil, S. Yusup and U. Rashid, *Fuel*, 2011, **90**, 2343–2345.
82. T. A. Isbell, *Grasas y Aceites*, 2011, **62**, 8–20.
83. Interactive European Network for Industrial Crops and Applications (IENIC(A)). Biolubricants Market Data Sheet. (2004), http://ienica. net/marketdatasheets/biolubricantsmds.pdf.
84. G. Booth, D. E. Delatte and S. F. Thames, *Ind. Crops Prod.*, 2007, **25**, 257–265.
85. E. Milchert and A. Smagowicz, *JAOCS*, 2009, **86**, 1227–1233.
86. Y. Xia and R. C. Larock, *Green Chem.*, 2010, **12**, 1893–1909.
87. J. B. Woerfel, *Practical Handbook of Soybean Processing and Utilization*, edited by D. R. Erickson, AOCS Press, Champaign IL, 1995, "Soybean oil processing by products and their utilization", pp. 297–313.
88. N. O. V. Sonntag, *Fatty Acids, utilization and disposal*, in *Edible Fats and Oils Processing: Basic Principles and Modern Practices*, edited by E. D. Erickson, AOCS Press Champaign IL, 1990, pp. 406–412.

89. (a) F. W. Keith Jr., F. E. Blachly and F. S. Sadler, *JAOCS*, 1954, **31**, 298–302; (b) M. K. Dowd, *JAOCS*, 1996, **73**, 1287–1295.
90. M. Hajek and F. Skopal, *Bioresour. Technol.*, 2010, **101**, 3242–3245.
91. A. Johansson, *Biomass*, 1982, **2**, 103–113.
92. A. J. Dijkstra and J. Segers, *The Lipid Handbook with CD-ROM*, CRC Press, 2007, "Production and refining of oils and fats", 3.6, 141–263. 3rd edn CRC Press, edited by F. D. Gunstone, J. L. Harwood, and A. J. Dijkstra.
93. J. B. Woerfel, *JAOCS*, 1983, **60**(2), 310–313.
94. Y. Ali, M. A. Hanna and S. L. Cuppett, *JAOCS*, 1995, **72**(12), 1557–1564.
95. L. Herrero-Dávila, *"Adding value to biodiesel glycerine and food waste"*, PhD thesis, University of York, September 2011.
96. R. Areski, Method for purification of glycerol, EP, 2007, 1978009.
97. M. J. Dumont and S. S. Narine, *Food Res. Int.*, 2007, **40**, 957–974.
98. C. Echim, R. Verhe, W. De Greyt and C. Stevens, *Energy Environ. Sci.*, 2009, **2**, 1131–1141.
99. Available online at: http://arizonachemical.com/en/Sustainability/ [Accessed: 26.04.11].
100. M. Carmona, J. Valverde and A. Perez, *J. Chem. Technol. Biotechnol.*, 2008, **84**, 738–744.
101. G. Leffingwell and M. Lesser, *Glycerin, its industrial and comercial applications*, Chemical Publishing Company, Brooklyn, NY, 1945, Chapter 1.
102. UK Environment Agency & WRAP, Biodiesel; Quality Protocol, June 2011. Available online at: http://www.environmentagency.gov.uk/static/documents/Leisure/090612_Biodiesel_QP_V5_final.pdf [Accessed: 11.08.11].
103. Available online at: http://www.novaol.it/novaol/cms/index.htmL [Accessed: 23.05.13].
104. GIA Industrial Analysis, Global *GlycerineMarket to Reach 4.4 billion Pounds by 2015*. Available online at: http://www.prweb.com/releases/glycerin_natural/oleo_chemicals/prweb4714434.htm [Accessed: 23.08.11].
105. Available online at: http://www.salisonline.org/market-research/biorenewable-chemicals-world-market/ [Accessed: 23.08.11].
106. (a) L. Ott, M. Bicker and H. Vogel, *Green Chem.*, 2006, **8**, 214–220; (b) *Biodiesel 2020: A Global Market Survey*, 2nd edition, 2008.
107. F. O. Licht's, World Ethanol and Biofuels Report; "Dark clouds on the Horizon for German Biodiesel sector", F.O. Licht GmbHF, 2007, 8, 6.
108. "Efficient electro-pressure technology for desalinating liquids (EET corporation)" Available online at: http://www.eetcorp.com/heepm/glycerine.htm. plus distillation, [Accessed: 23.08.11].
109. M. Pagliaro, R. Ciriminna, H. Kimura, M. Rossi and C. Della Pina, *Angew. Chem. Int. Ed.*, 2007, **24**, 4434–4440.
110. Official Journal of the European Communities. 2000, **226**, p. 3.
111. E. Ahna, M. Koncarb, M. Mittelbach and R. Marra, *Sep. Sci. Technol.*, 1995, **30**(7–9), 2021–2033.
112. Ms. S. Evans at BRENNTAG Group Ltd. Personal communication.

113. (a) V. L. Budarin, P. S. Shuttleworth, J. R. Dodson, A. J. Hunt, B. Lanigan, R. Marriott, K. J. Milkowski, A. J. Wilson, S. W. Breeden, J. Fan, E. H. K. Sin and J. H. Clark, *Energy Environ. Sci.*, 2011, **4**, 471–479; (b) V. L. Budarin, J. H. Clark, B. A. Lanigan, P. Shuttleworth, S. W. Breeden, A. J. Wilson, D. J. Macquarrie, K. Milkowski, J. Jones, T. Bridgeman and A. Ross, *Bioresour. Technol.*, 2009, **100**, 6064–6068.

114. Available online at: http://www.aquafuelresearch.com/glycerine-chp. htmL [Accessed: 24.06.11].

115. N. Kolesárová, M. Hutňan, V. Špalkov, Š. R. Kuffa and I. Bodík, *Chem.Papers*, 2011, **65**(4), 447–453.

116. N. Abatzoglou and S. Boivin, *Biofuels, Bioprod. Biorefin.*, 2009, **3**(1), 42–71.

117. M. S. Fountoulakis, I. Petousi and T. Manios, *Waste Manag*, 2010, **30**(10), 1849–1853.

118. S. S. Yazdani and R. Gonzalez, *Curr. Opin. Biotechnol.*, 2007, **18**, 213–219.

119. Available online at: http://www.bio-amber.com/succinic_acid.htmL, [Accessed: 24.07.11].

120. P. C. Lee, W. G. Lee, S. Y. Lee and H. A. N. Chang, *Biotechnol. Bioeng.*, 2001, **72**, 41–8.

121. E. Stoermer, S. Ledwell, A. Gu and R. Keeling, *Water Environ. Technol.*, 2009, **21**(9), 84–89.

122. Meet ANITAt the ammonia eater, Available online at: http://www.veoliawaterst.co.uk/medias/press/2012-07-16,Anita_Mox.htm [Accessed: 21.10.12].

123. C. Cherchi, A. Onnis-Hayden, and A. Z. Gu, Water Environment Federation Technical Exhibition and Conference, (WEFTEC), (19), 3149–3167, Chicago 2008.

124. (a) H. H. Moorer and C. M. Anderegg, Cement grinding aid and set retarder, US 4204877 (1980); (b) L. A. Jardine, J. H. Cheung, and W. M. Freitas, High early strength cement and additives and methods to make the same, US 6641661, (2003).

125. M. Rossi, C. Della Pina, M. Pagliaro, R. Ciriminna and P. Forni, *ChemSusChem*, 2008, **1**, 1196.

126. D. G. Rodgers, *Phenolic resin product and method of manufacturing a phenolic resin product*, WO 2007061903 A1, (2007).

127. Y. Gu and F. Jérôme, *Green Chem.*, 2010, **7**, 1127–1138.

128. (a) T. Werpy and G. Petersen, *Top Value Added Chemicals from Biomass*, U.S. Renewable Energy Laboratory (NREL), Oak Ridge, TN, USA, 2004; (b) J. J. Bozell and G. Petersen, *Green Chem.*, 2010, **12**, 539–554.

129. (a) D. Schreck, W. J. Kruper, R. D. Varjian, M. E. Jones, R. M. Campbell, K. Kearns, B. D. Hook, J. R. Briggs, and J. G. Hippler, Conversion of multihydroxylated aliphatic hydrocarbon or ester thereof to a chlorohydrin, WO 2006020234, (2006); (b) P. Krafft, C. Franck, I. De Andolenko and R. Veyrac, Process for the manufacture of dichloropropanol by chlorination of glycerol, WO 2007054505, (2007).

130. Int. patent WO2007010299, 2007, U.K. patent GB0514593, 2005.
131. S. Cassel, C. Debaig, T. Benvegnu, P. Chaimbault, M. Lafosse, D. Plusquellec and P. Rollin, *Eur. J. Org. Chem.*, 2001, **5**, 875–896.
132. J. A. Melero, R. Van Grieken, G. Morales and M. Paniagua, *Energy and Fuels*, 2007, **21**, 1782–1791.
133. Prof. G. J. Suppes awarded with 2006 Presidential Green Chemistry Challenge Awards: Available online at: http://www.epa.gov/green-chemistry/pubs/pgcc/winers/aa06.htmL [Accessed: 24.06.11].
134. J. Delgado, Procedure to obtain biodiesel fuel with improved properties at low temperature, US 0167681, (2003).
135. J. W. Yoo and Z. Mouloungui, *Stud. Surf. Sci. Catal.*, 2003, **146**, 757–760.
136. B. Katryniok, S. Paul, V. Bellière-Baca, P. Rey and F. Dumeignil, *Green Chem.*, 2010, **12**, 2079–2098.
137. (a) A. Takanori, Masayuki, Method for producing acrolein and method for producing acrylic acid, JP 2009292773, (2009); (b) A. Takanori and Y. Masayuki, Method for producing acrolein and method for producing acrylic acid, JP 2009292774, (2009).
138. N. Suzuki, Production method of acrolein, JP 2006290815, (2006).
139. (a) A. Neher, Process for the preparation of acrolein, DE 4238493, (1994); (b) A. Neher, Process for the production of acrolein, US 5387720, (1995).
140. S. Chai, *Green Chem.*, 2007, **9**, 1130.
141. A. Alhanash, E. F. Kozhevnikova and I. V. Kozhevnikov, *Appl. Catal. A*, 2010, **378**, 11–18.
142. A. Corma, G. W. Huber, L. Sauvanaud and P. O'Connor, *J. Catal.*, 2008, **257**, 163–171.
143. M. R. Nimlos, S. J. Blanksby, X Qian, M. E. Himmel and D. K. Johnson, *J. Phys. Chem. A*, 2006, **110**, 6145–6156.
144. W. Suprun, Mi. Lutecki, T. Haber and H. Papp, *J. Mol. Catal. A: Chem*, 2009, **309**, 71–78.
145. E. Tsukuda, *Catal. Commun.*, 2007, **8**, 1349.
146. W. Sun, J. Liu, X. Chu, C. Zhang and C. Liu, *J. Mol. Struct.: THEOCHEM.*, 2010, **942**, 38–42.
147. Y. Liu, H. Tüysüz, C. J. Jia, M. Schwickardi, R. Rinaldi, A. H. Lu, W. Schmidt and F. Schüth, *Chem. Commun.*, 2010, **46**, 1238–1240.
148. S. Queste, P. Bauduin, D. Touraud, W. Kunz and J.-M. Aubry, *Green Chem.*, 2006, **8**, 822–830.
149. (a) J. Barrault, F. Jerome and Y. Pouilloux, *Lipid Technol.*, 2005, **17**, 131–135; (b) Solvay International, *Polyglycerols, Versatile Ingredients for Personal Care*, November 2004; V. Plasman, T. Caulier and N. Boulos, *Plastic Additives & Compounding*, March/April 2005, Elsevier Ltd., ISSNI-464-39IX/05.
150. H. Gers-Barlag and A. Muller, Suncreen preparation containing surface active mono orolygoglyceryl compounds, water soluble UV-filter substance and, if desirered, inorganic micropigments, US Patent, US0072723 (A1), (2003).

151. (a) K. Klepáčová, D. Mravec, A. Kaszonyi and M. Bajus, *Appl. Catal. A*, 2007, **328**, 1–13; (b) K. Klepáčová, D. Mravec and M. Bajus, *Appl. Catal. A*, 2005, **294**, 141–147.

152. C. Marquez-Alvarez, E. Sastre and J. Pérez-Pariente, *Top. Catal*, 2004, **27**, 105–117.

153. J. Barrault, J. M. Clacens and Y. Pouilloux, *Top. Catal*, 2004, **27**, 137–142.

154. J. M. Clacens and J. Barrault, *Appl. Catal. A*, 2002, **227**, 181–190.

155. M. Ruppert, J. D. Meeldijk, B. W. M. Kuipers, B. H. Erné and B. M. Weckhuysen, *Chem. Eur. J*, 2008, **14**, 2016–2024.

156. P. Schuff-Werner, U. Muller, C. Unger, G. A. Nagel and H. Eibl, *Cryobiology*, 1988, **25**, 487–494.

157. T. W. Green and P. G. M. Wutts, *Protective Groups in Organic Synthesis*, *Wiley-Interscience*, New York, 1999, pp. 308–322.

158. E. García, M. Laca, E. Pérez, A. Garrido and J. Peinado, *Energy Fuels*, 2008, **22**, 4274–4280.

159. H. P. Staudinger and K. H. W. Tuerck, Methyl Glycerol production, CA 433404 A, (1946).

160. S. V. Koshchii, *Russ. J. Appl. Chem.*, 2002, **75**(9), 1434–1437.

161. S. Pariente, N. Tanchoux and F. Fajula, *Green Chem.*, 2008, **11**, 1256–1261.

162. I. Diaz, C. Marquez-Alvarez, F. Mohino, J. Prez-Pariente and E. Sastre, *Micropor. Mesopor. Mater.*, 2001, **44**, 295.

163. C. X. A. da Silva, V. L. C. Gonçalves and C. J. A. Mota, *Green Chem.*, 2009, **11**, 38–41.

164. R. Palkovits, I. Nieddu, C. A. Kruithof, R. J. M. Klein Gebbink and B. M. Weckhuysen, *Chem. Eur. J.*, 2008, **14**, 8995–9005.

165. F. Jerome and J. Barrault, *Eur. J. Lipid Sci. Technol.*, 2011, **113**, 118–134.

166. T. Watanabe, M. Shimizu, M. Sugiura, M. Sato, J. Kohori, N. Yamada and K. Nakanishi, *JAOCS*, 2003, **80**(12), 1201–1207.

167. (a) J. A. Ray, Process for improving hydraulic cement mixtures, GB 1593159, (1981); (b) S. Watanabe, *J. Oleo Sci.*, 2008, **57**(1), 1–10.

168. P. W. Brewster, C. R. Smith, and F. W. Gowland, Exxon Research Engineering Co., Glycerol ester as fuel economy additives, US 4683069, (1987).

169. B. Reinhard and M. Raymond, Textile lubricants containing glycerol triacetate, EP 0394802 (A1), (1990).

170. J. Delgado Puche, Process to obtain biodiesel fuel with improved properties at low temperature and comprising glycerine acetates, EU patent, EP1985684, A1, (2008).

171. N. Bremus, G. Dieckelmann, L. Jeromin, W. Rupilius, and H. Shuett, Process for the continuous production of triacetin, US 4381407 A, (1983).

172. J. E. Eastman, Etherified and esterified starch derivatives and processes for preparing same, US 4837314, (1989).

173. (a) A. Corma, S. Iborra, S. Miguel and J. Primo, *J. Catal.*, 1998, **173**, 315–321; (b) S. Bancquart, C. Vanhove, Y. Pouilloux and J. Barrault,

Appl. Catal., A, 2001, **218**, 1–11; (c) O. Koper, Y.-X. Li and K. J. Klabunde, *Chem. Mater.*, 1993, **5**, 500–505.

174. R. Luque, V. Budarin, J. H. Clark and D. J. Macquarrie, *Appl. Catal., B*, 2008, **82**, 157–162.

175. P. Ferreira, I. M. Fonseca, A. M. Ramos, J. Vital and J. E. Castanheiro, *Catal. Commun.*, 2009, **10**(5), 481–484.

176. X. Liao, Y. Zhu, S.-G Wang and Y. Lix, *Fuel Process. Technol.*, 2009, **90**, 988–993.

177. (a) H. E. Hoydonckx, D. E. De Vos, S. A. Chavan and P. Jacobs, *Top. Catal.*, 2004, **27**, 83; (b) W. D. Bossaert, D. E. De Vos, W. M. Van Rhijn, J. Bullen, P. J. Grobet and P. A. Jacobs, *J. Catal.*, 1999, **182**, 156.

178. S. Corma, S. Bee, Iborra and A. Velty, *J. Catal.*, 2005, **234**, 340.

179. F. Jerome, Y. Pouilloux and J. Barrault, *ChemSusChem*, 2008, **1**(7), 586–613.

180. I. Diaz, C. Marquez-Alvarez, F. Mohino, J. Pérez-Pariente and E. Sastre, *Micropor. Mesopor. Mater.*, 2001, **44**, 295.

181. F. Jerome and Joel Barrault, *Eur. J. Lipid Sci. Technol.*, 2011, **113**, 118–134.

182. G. J. Suppes, *Catalyst and method for improved glycerol reduction*, A presentation at the Bio-Futures Conference, Saskatoon (Canada), 16–17 October, 2006.

183. M. A. Dasaria, P. P. Kiatsimkula, W. R. Sutterlin and G. J. Suppes, *Appl. Catal., A*, 2005, **281**, 225–231.

184. J. Henkelmann, M. Becker, J. Buerkle, P. Wahl, and G. Theis, process for the preparation of 1,2-propanediol, WO 2007099161, (2007).

185. C-W. Chiu, M. A. Dasari, W. R. Sutterlin and G. J. Suppes, *Ind. Eng. Chem. Res.*, 2006, **45**, 791–795.

186. R. R. Soares, D. A. Simonetti and J. A. Dumesic, *Angew. Chem. Int. Ed.*, 2006, **45**, 3982–3985.

187. S. Shelley, *Chem. Eng. Prog.*, 2007, **103**(8), 6–10.

188. (a) M. W. M. Tuck and S. N. Tilley, Process for the production of 1,2-propanediol, US 2007149830 (A1), (2007); (b) M. W. M. Tuck, Process for the production of a hydroxilic compound, MXPA94009224 (A), (2005).

189. C. Fang, C. Jing, X. Chungu, K. Haixiao, X. Xinzhi, T. Jin, and L. Xuemei, Method for producing 1,2-propyleneglycol using bio-based glycerol, US 0156866 (A1), (2009).

190. (a) C. A. McKinzie and J. H. Stocker, *J. Org. Chem.*, 1955, **20**, 1695; (b) F. Roessner, P. Adryan, R. W. Fischer, R. A. Rakoczy, Future Feedstocks for Fuels and Chemicals, Procedings of the DGMK Conference, Berlin, Germany, September 2008.

191. Y. Shi, W. Dayoub, G. Rong Chenb and M. Lemaire, *Green Chem.*, 2010, **12**, 2189–2195.

192. C. X. A. da Silva, V. L. C. Gonçalves and C. J. A. Mota, *Green Chem.*, 2009, **11**, 38–41.

193. E. García, M. Laca, E. Pérez, A. Garrido and J. Peinado, *Energy & Fuels*, 2008, **22**, 4274–4280.

194. (a) Available online at: http://www.freedoniagroup.com/Polyurethane. html [Accessed: 24.06.11]; (b) A. Dibenedetto, A. Angelini, M. Aresta, J. Ethiraj, C. Fragale and F. Nocito, *Tetrahedron*, 2011, **67**(6), 1308–1313.
195. R. A. Grey, Preparation of cyclic carbonates using alkylammonium and tertiary amine catalysts, US5091543, (1992).
196. A.-C. Simão, B. Lynikaite-Pukleviciene, C. Rousseau and A. Tatibouët, *Lett. Org. Chem.*, 2006, **3**(10), 744–748.
197. (a) R. García, M. Besson and P. Gallezot, *Appl. Catal., A*, 1995, **127**, 165–176; (b) S. Carrettin, P. McMorn, P. Johnston, K. Griffin, C. J. Kiely and G. J. Hutchings, *Chem. Phys.*, 2003, **5**, 1329–1336.
198. Available online at: http://www.eetcorp.com/heepm/glycerine.htm, [Accessed: 24.05.13].

CHAPTER 7

Industrial Use of Oil Cakes for Material Applications

ANTOINE ROUILLY*[a,b] AND CARLOS VACA-GARCIA[a,b]

[a] Université de Toulouse INP-ENSIACET, LCA (Laboratoire de Chimie Agro-Industrielle), F-31030 Toulouse, France; [b] INRA, UMR 1010 CAI, F-31030 Toulouse, France
*Email: Antoine.Rouilly@ensiacet.fr

7.1   Introduction

Nowadays, the main industrial outlet for oil cakes is animal feed. The main reason for this is its high content of protein. But in this field the competition is unfair; soy meal is produced in much higher volume (63% of the world production of oil cakes, 61% of the European production[1]) and has the best nutritional value. For the other crops, especially sunflower and rapeseed, oil cakes are still considered as a byproduct. They are available in huge amounts (European production of sunflower and rapeseed oil cakes were, respectively, 3.4 and 12.0 Mt in 2011[1]) and at a reasonable price.

For these reasons, high protein content and low price, oil cakes are interesting candidates for the production of biomaterials. Furthermore, in industrial fallow lands a complete use of oilseed crops could be considered for nonfood applications, while keeping on the land enough biomass for soil improvement.

To assess the scientific, technological, and economical feasibility of the transformation of oil cakes into biomaterials, we will present first the plastic properties of oilseed proteins, then the thermomechanical transformation (compounding and injection moulding) of sunflower oil cake, and lastly an economic and industrial evaluation of oil cake-based materials.

RSC Green Chemistry No. 24
The Economic Utilisation of Food Co-Products
Edited by Abbas Kazmi and Peter Shuttleworth
© The Royal Society of Chemistry 2013
Published by the Royal Society of Chemistry, www.rsc.org

7.2 Plastic Properties of Oilseed Proteins

7.2.1 Extraction of Oilseed Proteins

Extraction processes of proteins from sunflower and rapeseed oil cakes were kept confidential until the last decade. But nowadays many scientific studies are dedicated to this purpose, while the nonfood industrial use of polypeptides gains interest for various applications such as surfactants, glues, supported antioxidants, or encapsulation compounds.[2]

The main concerns regarding the protein extraction are:

- the control of protein structure and degradation during the separation process in order to maintain protein solubility and functional properties;[3,4]
- the interactions of proteins with phenolic compounds, which can be an asset for antioxidant and antimicrobial properties[5] and a major drawback for the colouration of the extract;[6]
- the integration of the process at industrial scale[7] with the oil extraction process.[8]

7.2.2 Complex Structures and Interactions

Soybean proteins have been known and used for a long time due to their plastic properties.[9] Formerly, it was more their chemical reactivity that was exploited to make thermosetting resins. Among the 22 amino acids constituting the complex structure of proteins, many reactive moieties are available to create covalent crosslinking (Figure 7.1). This can occur by self-crosslinking or

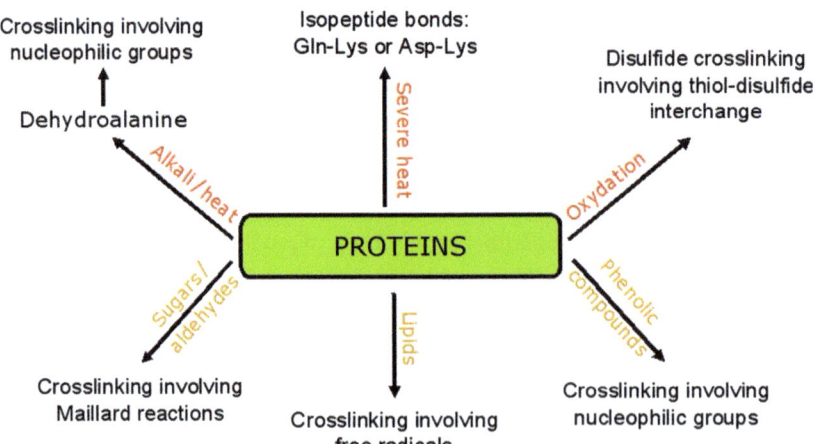

Figure 7.1   Major types of covalent crosslinking occurring with proteins. Adapted from reference.[10]

with the participation of other biomolecules like carbohydrates and lipids.[10] This reactivity was the starting point of the nonfood use of protein and has been widely used in the preparation of adhesives.[11]

The physicochemical properties of polymers (biopolymers being one example among others) are also driven by the noncovalent bonds between the chains. Also, given the large variety of monomers in peptides, almost all kinds of interactions are possible (Figure 7.2). These interactions are responsible for the intricate macromolecular organisation of proteins and maintain secondary, tertiary, and quaternary structures.[12,13]

Protein material is therefore organised in a complex manner. The investigation of a potential thermomechanical transformation by making the best use of their thermoplastic properties needs to consider the macromolecular structure modifications.

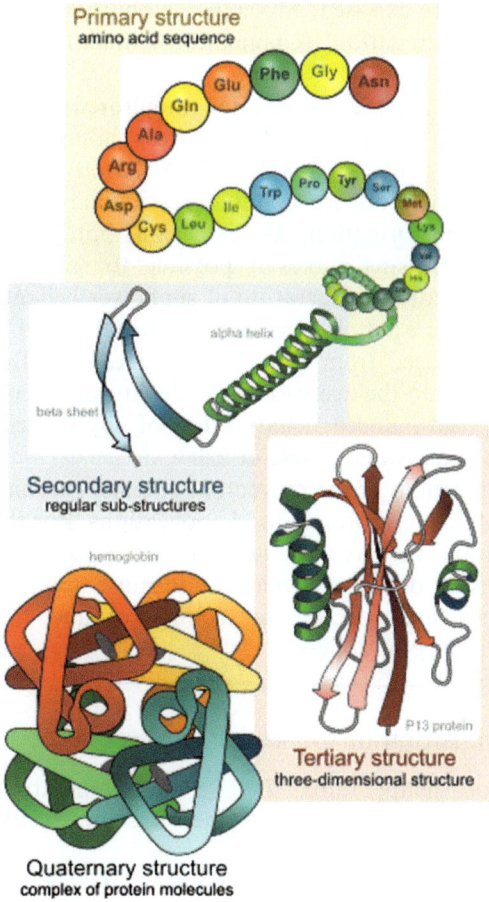

Figure 7.2 Macromolecular organisation of proteins.

7.2.3   Thermal Denaturation of Proteins

Protein denaturation is commonly defined as any noncovalent change in the structure of a protein. This change may alter the secondary, tertiary, or quaternary structure of the molecules. Modifications in the structure of proteins can be caused by many factors, *e.g.*, heat, pH, dielectric constant and ionic strength.[14]

When proteins are exposed to increasing temperature, losses of solubility or enzymatic activity occur over a fairly narrow range. Depending upon the considered protein and the severity of the heating, these changes may be reversible or irreversible. As the temperature is increased, a number of bonds in the protein molecule are weakened. The first affected are the long-range interactions that are necessary for the configuration of the tertiary structure. As these bonds are first weakened, and then broken, the protein becomes more flexible. As heating continues, some of the cooperative hydrogen bonds that stabilise the helical structure will begin to break and hydrophobic groups are exposed. Molecular mobility and high temperature result in protein aggregation (coagulation), making the denaturation irreversible.

The denaturation phenomenon is often monitored by DSC analysis of a protein solution or suspension. The corresponding endothermic peak, which depends on the conformational state of the protein, is found around 70–90 °C.[15] Under low moisture conditions, similar to those used for a thermomechanical transformation (*i.e.* extrusion, hot pressing), the use of pressure-resistant DSC pans makes it possible to study the denaturation process.[16] The denaturation temperature of sunflower helianthinins, the main protein in sunflower, for example, decreased from 190 to 120 °C as the oil cake moisture content increased from 0 to 30% (Figure 7.3). Exothermic aggregation was then related to the decrease of the denaturation enthalpy but, more recently, work on heat-induced denaturation of soy proteins stated that a specific exothermic peak is observable at 220 °C.[17]

Controlled denaturation, or the prevention of denaturation, is an important aspect in the development of protein food applications. For the use of proteins in nonfood applications such as surfactants, adhesives, coatings or plastics, it is accepted that a certain degree of denaturation must occur in order to make proteins processable, and to reach the required product properties such as strength, water resistance or adhesion.[18]

7.2.4   Glass Transition of Oilseed Proteins

The glass-transition temperature, T_g, is the temperature at which an amorphous solid, such as glass or a polymer, becomes brittle on cooling, or softens on heating.

In the 1980s, with the intense development of theories on the glassy state of foods,[19] almost all biopolymers have been tested to assess their T_g. Among oilseed proteins, only soy globulins were considered.[20] But for many

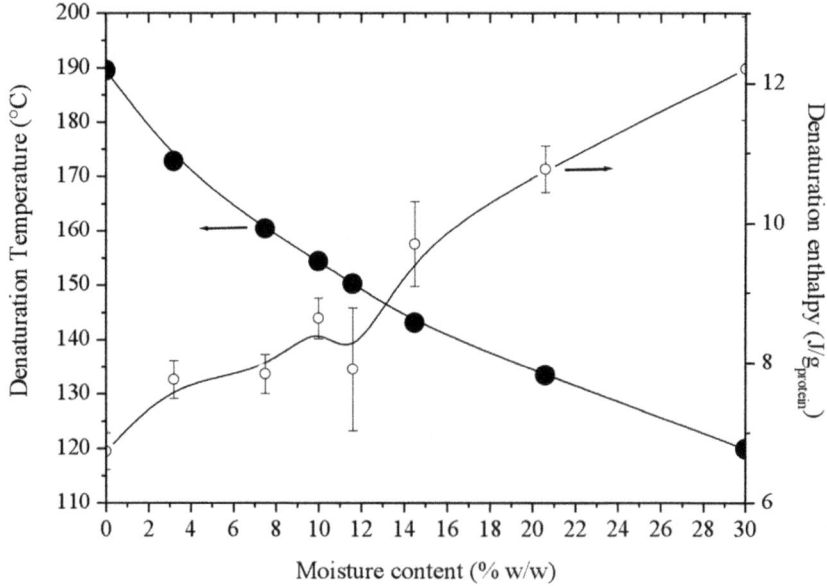

Figure 7.3 Denaturation temperature and enthalpy of sunflower proteins according to moisture content.[16]

reasons the glass-transition points of plant-based proteins are not easily detected:

- they are ordered in macromolecular structures (Figure 7.2) and should be considered as semicrystalline polymers;[21]
- thermal denaturation, considered as "fusion" of the crystallites, is irreversible and induces changes in chain mobility (exposure of concealed groups, crosslinkings, *etc.*);
- the strong dependence of T_g with moisture content makes the measurement difficult: state changes of adsorbed water, displacement of adsorbed water with temperature increase;
- different transitions related to side-groups and the main backbone, often referred to as β and α transitions, respectively, are observable.[22]

Nevertheless, for protein isolates, which are often denatured by the extraction process, the glass transition has been studied. For sunflowers, for example, a classical behaviour towards moisture content has been reported and modeled;[23] T_g decreases from 180.8 °C to 5.3 °C when the protein moisture content increases from 0 to 26.1% w/w on a dry basis (Table 7.1). This kind of measurement has never been done on rapeseed.

An interesting review on the processing of proteins according to their thermal properties has been proposed by De Graaf[18] and is described in Figure 7.4. In order to denature a protein, the system defined by the temperature and the water content should be situated above the denaturation

Table 7.1 Evolution of glass-transition temperature of sunflower proteins according to their water content.[23]

R.H. (%)	Water Content (% DM)	T_g (°C)
–	0	180.8
8	3.05	129.5
33	5.95	113.1
43	7.40	111.9
54	9.03	77.8
62	11.67	68.4
75	13.47	46.1
80	15.73	39.6
85	17.09	33.8
92	26.12	5.3

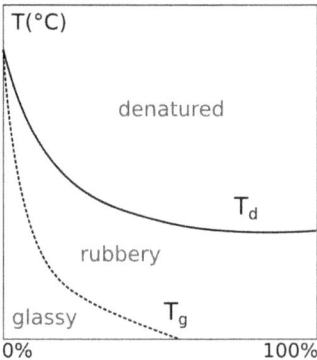

Figure 7.4 Dependence of denaturation temperature (TD, solid line) and glass transition temperature (TG, dashed line) of proteins on the water content. Adapted from reference.[18]

temperature (T_d) line, allowing both thermodynamic denaturation ($T > T_d$) and sufficient chain mobility ($T > T_g$). When processing is considered, this condition can be achieved in three ways:

- by increasing the water content, allowing processing at 'low' temperature;
- by increasing the temperature, which is particularly effective at low water content;
- by adding structure breakers/denaturants (*e.g.* urea), which reduce T_d at a given water content.

7.2.5 "Melt" Rheology of Oilseed Proteins

While extrusion of protein-based materials is of great interest for food applications,[24] few studies have been performed to characterise the rheological behaviour of low moisturised protein systems.[25] Torque rheometry experiments have been conducted on gluten[26] and soy protein isolate[27] showing the occurrence of shear-induced crosslinking. Concerning the capillary rheology, according to our knowledge, the only study has been performed on sunflower protein isolate.[28] It shows that below the denaturation temperature, sunflower protein has a shear thinning and pseudoplastic behaviour and the "melt" apparent viscosity follows the power-law model. Reduction of disulfide bonds with sodium sulfite greatly enhances the mixtures flowing characteristics (Figure 7.5).

As disulfide bonds are a key factor in the thermomechanical transformation of protein-based materials, further explanations seem necessary.

Cysteine bridges are the only nonpeptidic covalent bonds in proteins. Formed by the oxidation of cysteine-thiol groups, they contribute to the stabilisation of the three-dimensional structure of proteins. Cysteine bridges of sunflower albumins have been mapped[29] but those of the globulins have not yet been directly studied. As most of the dicotyledon globulins have similar structures,[30] their concentration must be close to those measured for linseed globulins, that is 61.4 µmol/g of proteins.[31] On the whole protein fraction, their concentration is probably close to that measured on the soybean protein isolate, approximately 70 µmol/g.[32] Their reversible reduction[33] frees the protein chains from native constraints, giving them more mobility (Figure 7.6). Many molecules have been tested but the cheapest and efficient, is sodium sulfite.

When denaturation is reached during processing, protein viscosity increases because of an association/coagulation phenomena. Injection moulding of

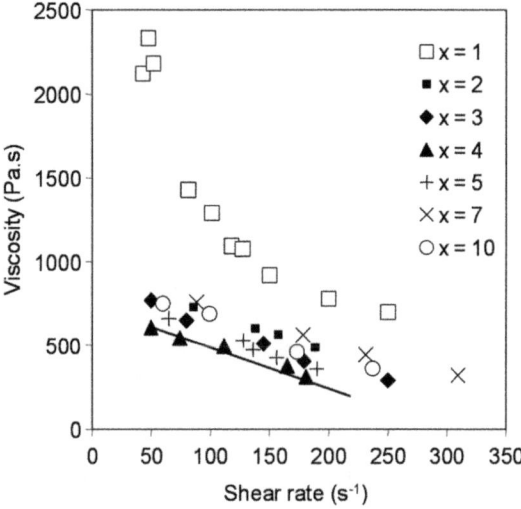

Figure 7.5 Effect of sodium sulfite content on the viscosity of mixtures of SFPI-glycerol-water-sodium sulfite = 100–30–30–x parts in weight. Temperature of the die = 120 °C.[27]

$$R\text{-}SS\text{-}R' + SO_3^{2-} \rightleftharpoons R\text{-}SSO_3^- + R'S^-$$

$$SO_3^{2-} + H^+ \rightleftharpoons HSO_3^- \quad (pK_a = 6{,}8)$$

$$RS^- + H^+ \rightleftharpoons RSH \quad (pK_a >> 8{,}5)$$

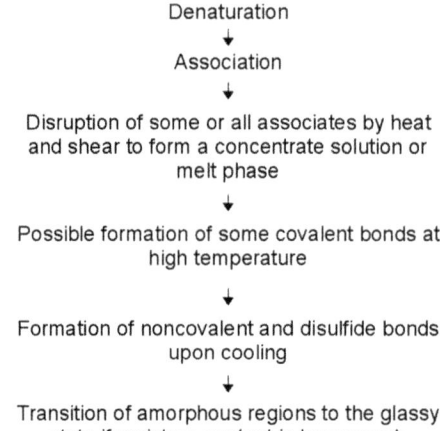

Figure 7.6 Reduction scheme of disulfide bonds by sulfite ion.[33]

Denaturation
↓
Association
↓
Disruption of some or all associates by heat and shear to form a concentrate solution or melt phase
↓
Possible formation of some covalent bonds at high temperature
↓
Formation of noncovalent and disulfide bonds upon cooling
↓
Transition of amorphous regions to the glassy state if moisture content is low enough

Figure 7.7 Protein structural change occurring during the extrusion process.[24]

sunflower protein isolate was only possible at processing temperatures below the denaturation temperature.[28]

To summarise the structural changes of protein during an extrusion process, Arêas has proposed an interesting "suspension model"[24] presented in Figure 7.7.

7.2.6 Oilseed Protein-Based Materials

Two research groups have intensively worked on oilseed protein-based materials: Jay-Lin Jane's group in Iowa State University and Françoise Silvestre's one in INP Toulouse, respectively, on soybean and sunflower.

Excluding wet forming by casting – as it is not industrially viable for nonfood applications – the main forming technology used to produce protein-based materials is the thermomoulding or hot-pressing of flat sheets. Applying high temperature and pressure surpasses the T_g and T_d and, as no flowing properties are required, coagulation/association phenomena is favored. The resulting materials have some mechanical properties ranging from highly resistant and brittle materials to highly plastic materials (elongation up to 300%) depending on the plasticiser nature and content.[34–38] They also exhibit some interesting properties towards water when compared to carbohydrate-based materials due to both coagulation and natural disulfide crosslinking.[37]

The use of continuous or semicontinuous processes, such as extrusion and injection moulding is more problematic due to classic aggregation following the denaturation phenomenon. Basically, three different ways have been studied to overcome this problem:

- the use of a coconstituent like starch,[39] a biodegradable polyester,[40–42] or lignocellulosic fibres;[43]
- processing at temperatures below T_d;[22,28,44]
- the addition of large amounts of plasticisers.[45]

7.2.7 Other Applications of Oilseed Proteins

Apart from their thermoplastic properties, many other properties of oilseed proteins have been investigated for industrial applications. Among them, and nonexhaustively, they can be used as binders in cellulosic paper[46] or in particleboards,[47,48] some hydrolysates have some interesting foaming and emulsifying properties,[49] or they may replace carbohydrates in encapsulation.[6]

7.3 Injection Moulding of Sunflower Oil Cake

Oilseed proteins possess interesting plastic properties but their complicated behaviour when submitted to shear and heat hinders their industrial development as a biopolymer. For thermomechanical processes (*i.e.* extrusion, injection moulding) complete coagulation has to be avoided and a way to succeed is to add some lignocellulosic fibres.

Oil cake is the crude byproduct of the oil extraction. It contains residues of the seed kernel (protein, cell walls, *etc.*) and the whole hull, which is constituted of lignocellulosic fibres. It is therefore an interesting substrate to illustrate the transformation of a complete industrial byproduct into a "plastic" biocomposite.

We can note that only soy meal and sunflower oil cake have been tested for protein-based materials.[11,50,51] In the case of soy, protein concentrate or isolate are more likely used since the industrial protein extraction process yields an oil cake poor in fibres. Other oilseeds currently used for the production of oil such as rapeseed or linseed have not been considered.

7.3.1 Structure and Composition of Sunflower Oil Cake (SFOC)

7.3.1.1 Oil-Extraction Process

The complete analysis of SFOC structure and composition has to start with the description of the industrial oil extraction process as it involves a large numbers of steps that could cause drastic changes in the physical and chemical properties of its constituents. The main steps of the industrial extraction and their consequences are summarised in Table 7.2.

Table 7.2 Processes and biological consequences during the main steps of the industrial extraction of sunflower oil.

Step	Process	Consequences
Flaking	Plain flaking rolls	Formation of 0.3 mm thick flakes
		Hull break down
		Contact area increase
Cooking	Steam cooking (80 °C)	Setting of flakes moisture content (3–5%)
	Drying (110 °C)	Decrease of oil viscosity
		Increase of cell-wall plasticity
		Protein denaturation
		Sterilisation
Pressing	Shear/compression	Cell-wall break down
		Oil (and coproducts) expression
Extraction	Solvent (hexane)	Oil solubilisation
Desolvation	Steam distillation (80 °C)	Protein coagulation
	Steam toasting (110–115 °C)	Formation of phenolic/proteic aggregates
	Drying (95–100 °C)	Setting of moisture content for preservation

Table 7.3 SFOC composition.[52]

Components	Methods	SFOC1	SFOC2	SFOC3
Moisture content (%DM)	NF V 03-903	9.9	11.0	10.3
Proteins	NF V 18-100	35.0	29.4	31.5
Cellulose	ADF-NDF	18.3	24.2	25.1
Lignin		9.1	8.1	8.0
Hemicelluloses		10.9	15.4	12.5
Lipids	NF ISO 0734-1	1.5	1.3	3.2
Chlorogenic acid	UV spectroscopy (Folin–Ciocalteu reagent)	2.7	2.9	–
Phenolic compounds	Estimated	4.5	4.9	–
Water-soluble components	Boiled water	24.9	22.4	21.2
Ashes	NF V 03-922	7.2	7.0	7.1

7.3.1.2 Composition of Sunflower Oil Cake

The composition of three SFOC batches is described in a recent paper[52] and is shown in Table 7.3. The SFOC composition is considerably stable even if intrinsic and extrinsic factors (genetic, climate, soil, and extraction process) induce small changes in their composition. The main part of the SFOC is made of cell-wall components such as lignocellulosic fibres (around 40%/dry matter). The second major part is represented by protein fractions (around 30%/DM). The other 30% is made up of many other components that contribute to the matrix or to the fillers, many of which are water soluble.

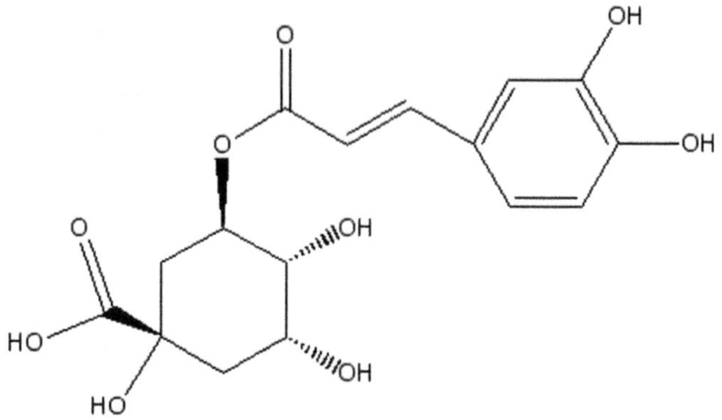

Figure 7.8 Chemical structure of the chlorogenic acid.

Table 7.4 Protein distribution according to the classification of Osborne and Campbell.[52]

Protein class (%/total proteins)	Solvents	SFC1	SFC2	SFC3
Globulins	1 m Na Cl	28.9	29.2	34.4
Glutelins	NaOH, pH 11.0	17.3	25.3	23.2
Albumins	Distilled water, pH 6.5	0.7	0.2	0.2
Prolamins	Aqueous ethanol, 70%	2.1	3.2	3.0
Insoluble proteins		51.0	42.1	39.2

The analysis reveals a significant amount of phenolic acid: the chlorogenic acid (Figure 7.8). Most of the time found polymerised, this compound is known as a complexing agent of protein.[53,54] Their polyfunctional structure gives them the ability to bind to other molecules by all kinds of secondary interactions (hydrogen, hydrophobic and ionic). Under oxidative conditions they form a very reactive quinone and induce covalent crosslinking. The high amount of polyphenolic compounds is responsible for the dark green colouration of sunflower protein isolate and is often considered as a major drawback (along with its laxative properties) for the feeding value of SFOC.

The protein composition of SFOC, determined by the selective extraction of proteins as described by Osborne and Campbell, is presented in Table 7.4. There are few results available on the protein composition of SFOC in the literature. Most of the results are given for sunflower dehulled cake meal or defatted meal, rather than for industrial SFOC. The very low content of recoverable albumins (<1%) and the high insoluble protein content (>35%) is related to the severity of the industrial treatment.

7.3.1.3 Structural Analysis of SFOC

Macroscopically, only hull fragments are distinguishable, they represent about 40% w/w of the SFOC (Figure 7.9). SEM analysis of these fragments revealed

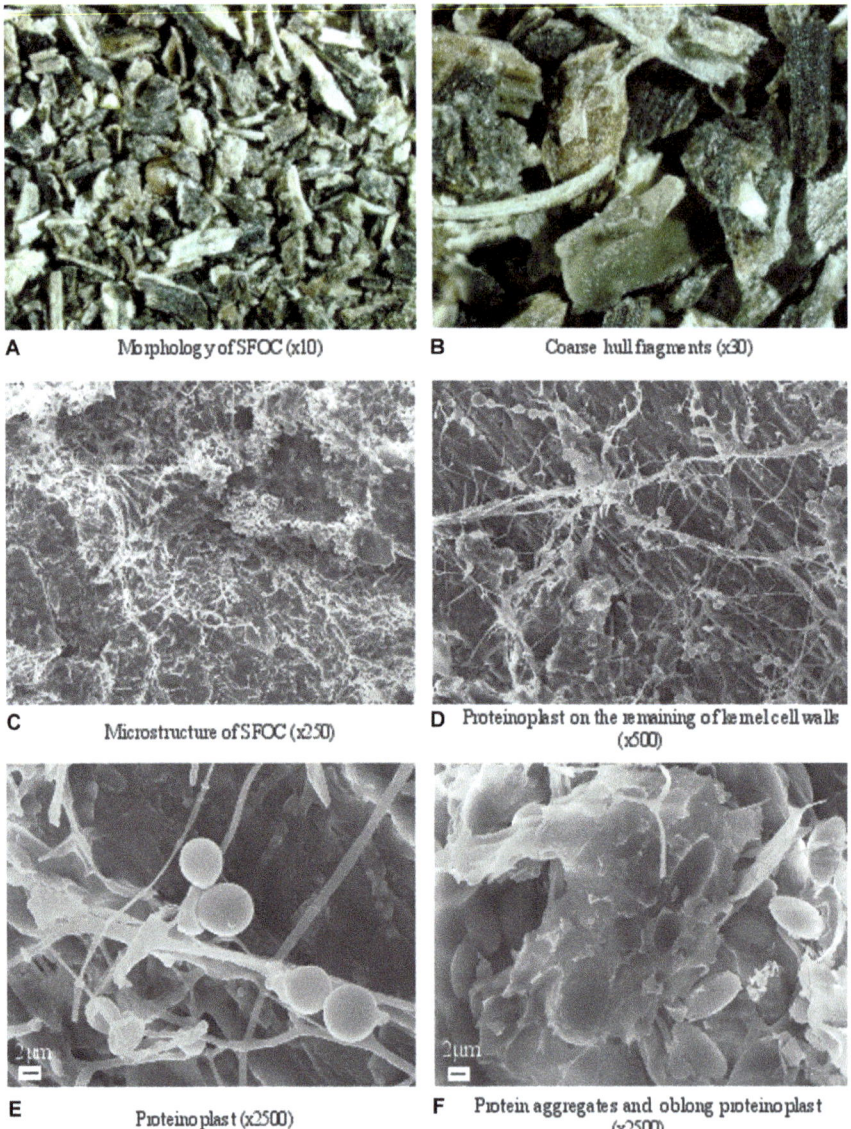

A   Morphology of SFOC (x10)   **B**   Coarse hull fragments (x30)

C   Microstructure of SFOC (x250)   **D**   Proteinoplast on the remaining of kernel cell walls (x500)

E   Proteinoplast (x2500)   **F**   Protein aggregates and oblong proteinoplast (x2500)

Figure 7.9   Light and electron micrographs of SFOC.

the remaining parts of the kernel deposed on the surface: fine cell-wall fibres, undefined clusters and intact proteinoplasts.

Sunflower proteins are considered denatured by oil extraction[55] as confirmed by the high amount of insoluble proteins. But a recent DSC study of the thermal denaturation of sunflower protein in the seed, in the oil cake, and in the isolate showed no enthalpy difference.[52]

At a higher magnitude, two kinds of corpuscles appear. Isolated proteinoplast have a spherical shape from a diameter between 2 and 6 μm while some other organoplasts have an oblong shape and seem to be aggregated. These aggregates could correspond to the insoluble protein found in high amount (Table 7.4).

Oil extraction does not affect the thermal stability of the corpuscular proteins of sunflower seed but is responsible for the aggregation into a large extent of protein with polyphenolic compounds.

7.4 Twin-Screw Extrusion of SFOC

Twin-screw extrusion is a versatile tool well known for:

- the defibration or pulping of lignocellulosic fibres;[56]
- the texturation of protein;[24]
- the compounding of thermoplastic materials: mixing additives, pelletising, *etc.*

Twin-screw extrusion could then in a one step process, plasticise the protein fraction and defibrate the husk fragments of SFOC.[57] A classical screw profile and the operating conditions used are shown in Figure 7.10.

Using this kind of basic process: only water is needed (the moisture content is set at 30%), the temperature is limited at 100 °C and specific mechanical energy is kept under 300 W h/kg, SFOC microstructure is completely changed.

The small and hard particles of SFOC (Figure 7.9) are transformed during extrusion into small and soft fibre aggregates (Figure 7.11). The bulk relative density of the extruded sunflower oil cake (ESFOC) decreases from 0.45 after the grinding of the granules to 0.29 g cm^{-3} after extrusion. The softness of the ESFOC can be attributed to the moistening of the initial mixture and to the defibration of the fibrous fragments of hull.

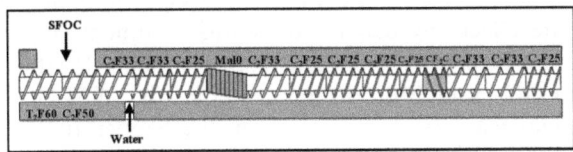

Parameters	Values
Barrel temperatures (°C)	100
Screw speed (t/min)	200
Solid feed rate (kg/h)	22.3
Liquid feed rate (kg/h)	6.5
Output (kg/h)	26.2
Output moisture content (%)	21.2
Specific mechanical energy (W.h/kg)	277.6

Figure 7.10 Twin-screw extruder configuration and operating conditions for the processing of SFOC.[57]

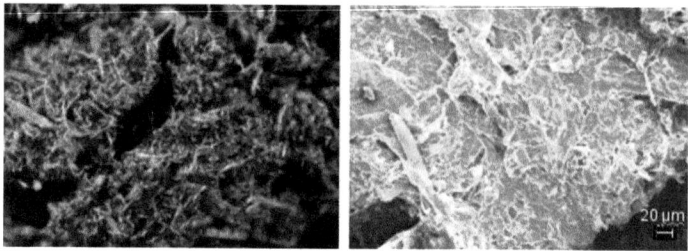

Figure 7.11   Light (left × 30) and electron (right × 250) micrographs of extruded SFOC.

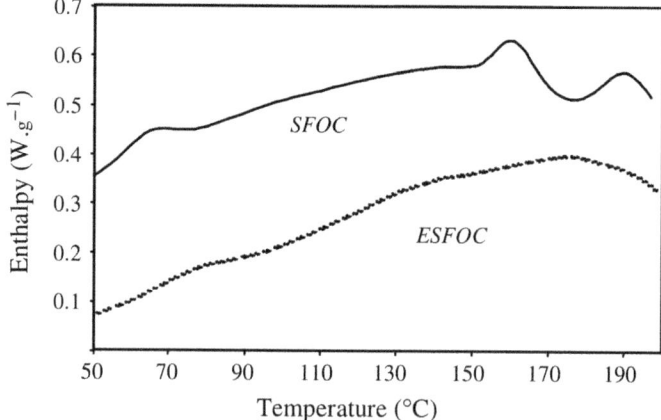

Figure 7.12   DSC thermograms (endo up) of SFOC and ESFOC. Samples equili-
brated at 60%RH and 25 °C.

The apparition of fibrous aggregates could be credibly connected to a
complex process of association of the noncellulosic compounds of the mixture.
The globulins are effectively denatured by this treatment (Figure 7.4). From
DSC analyses, the peak corresponding to the globulin denaturation (around
160 °C for SFOC samples) disappears on the ESFOC thermogram
(Figure 7.12). This was observed even when the temperature of the extruder
(Figure 7.10) was lower than the temperature of denaturation measured in DSC
for a sample with 30% moisture content.[16] Two reasons could explain this
phenomenon: the local temperature rise in zones of higher stress[58] and the
influence of the shear.

Besides, further to the denaturation, temperature promotes a set of
association reactions involving the other protein fractions, the phenolic
compounds, possibly the ligneous byproducts released by the defibration of the
husk fragments and sugars (Maillard reactions).[24] This could explain the
apparent homogeneity observed in the scanning electron micrographs
(Figure 7.11). The protein corpuscles of the raw sunflower oil cake (Figure 7.9)
disappear after the treatment and the fibres are embedded in a continuous
matrix.

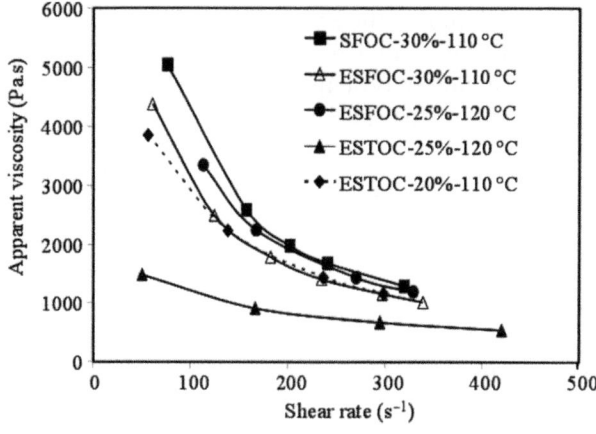

	K (Pa.s^m)	m	R^2
SFOC-30%-110°C	318640	0.04	0.9994
ESFOC-30%-110°C	147843	0.15	0.9988
ESFOC-25%-120°C	310958	0.04	0.9996
ESTOC-25%-120°C	9145	0.54	0.9923
ESTOC-20%-130°C	70097	0.29	0.9917

Figure 7.13   Apparent viscosity of SFOC, ESFOC and ESTOC samples at different moisture contents and temperatures and corresponding power law coefficients.[57]

The structural modification of SFOC results in a substantial reduction in its apparent viscosity (Figure 7.13). Thus, after the defibration of the husk fragments and the "fusion" of the protein corpuscles, the ESFOC viscosity at 25%MC was equivalent to the SFOC viscosity at 30%MC. And "melt" viscosity is the key parameter for the injection-moulding process. First, the flow of the mixture being easier, the mould filling is better; and secondly, a decrease of the moisture content of the compound results in a lower deformation of moulded samples upon drying.

Like starch,[59] hydrated sunflower oil cake was shear thinned and its viscosity followed a power-law model. The improvement of the rheological behaviour of the SFOC after extrusion is demonstrated by the significant decrease of the consistency coefficient K and the increase of the pseudoplasticity index m (Figure 7.13). But this index remained low in comparison with those obtained for starch,[60] possibly explained by the strong interactions within the polypeptide entanglements.[20] Despite the higher temperatures, the decrease of the moisture content led to a diminution of the m index.

The addition of a reducing agent (here sodium sulfite) greatly enhances the rheology of the ESFOC as was described in the case of sunflower protein isolate.

Although the reduction of disulfide bridges also involves a decrease in the temperature and the denaturation enthalpy,[31] the treatment was only employed on the ESFOC. The evolution of the ESFOC viscosity according to the amount of reducing agent is similar to that observed in a previous study of the rheological properties of sunflower protein isolate.[28] The increase of the sodium sulfite ratio with regard to the mass of protein from 0 to 5% results in a progressive decrease of the apparent ESFOC viscosity. Then, when the addition is increased above 5%, the viscosity does not decrease further (Figure 7.5), and even increased when a low shear stress rate was applied to the mixture. This result is attributed to the consequence of the structural changes in proteins during the chemical attack, however, the mechanism explaining this phenomenon has not yet been elucidated.

Extruded sunflower oil cake treated with 5% sodium sulfite (ESTOC) has an optimal rheological behaviour. The addition of the reducing agent leads to a large decrease in the consistency coefficient K and an increase of the pseudoplasticity index m (Figure 7.13). At a moisture content of 25% and a temperature of 120 °C, the coefficients of consistency and pseudoplasticity are respectively, 310 958 and 0.04 for the ESFOC and 9145 and 0.54 for the ESTOC. The rheological behaviour of ESTOC is particularly interesting and as such, it would be possible to decrease its moisture content further. Therefore, while the treatment *via* extrusion results in a 5% decrease of the moisture content of the cake, treatment with sodium sulfite reduces it more efficiently (Figure 7.13).

7.5 Injection Moulding of SFOC

Injection moulding of 100% natural biopolymers and biocomposites is not well documented. It still belongs mainly to industrial know-how, which stands as a major retarding issue for the development of agromaterials. Only two German papers[61,62] from the 1990s compare the operating conditions for the moulding of such materials. From these works and according our experience, particular issues have been raised and are summarised in Table 7.5. There are three main differences for their processing when compared to classic thermoplastic materials:

- the occurrence of adsorbed water: agrobased compounds should never be dried;
- the heat sensitivity of biopolymers: high shear, long residence time and high temperature should be avoided;
- the high viscosity of these materials: high injection pressure and wide gaps should be used.

For the processing of SFOC, injection moulding at different stages of the destructuring process has been studied. Raw SFOC with 30% moisture content can be formed by injection moulding. However, due to its lack of "plasticity", the use of a screw without a backflow stop valve is necessary. The mechanical constraint engendered by this valve resulted in a local temperature rise and

Table 7.5 Processing conditions for the injection moulding of SFOC-based materials.

Operating Conditions	Synthetic polymers	SFOC-based pellets
Screw L/D	High (28) for heat-resistant polymers and low (18) for heat sensitive polymers	Low (20) to decrease residency time
Compression ratio (τ_c)	Between 2 and 2.8 for most materials, high for powders	Between 1.8 and 2 as agropellets are dense and heat sensitive (self-heating and degassing)
Screw speed	Depending on applications	Average to avoid self-heating
Screw backpressure	Between 3.5 and 21 bar, high for viscous materials	Between 10 and 20 bar to avoid water vapour degassing voids
Injection speed	High to improve pace and decrease material viscosity	Low to control injection pressure
Injection pressure	Low (hundred of bar)	High as materials are more viscous (1500–2000 bar)

Table 7.6 Injection moulding conditions and mechanical properties in tension (σ_t : stress at break, E_y : Young modulus) and bending (σ_f : stress at break, E_f : bending modulus) of SFOC, ESFOC and ESTOC moulded specimens.[57]

	SFOC	*ESFOC*	*ESTOC*
Moisture Content (%)	30	25	20
T_{max} barrel (°C)	50	120	130
T mould (°C)	100	room	room
Average density	1.09	1.20	1.34
Flexural properties			
σ_{max} (MPa)	5	11.1 ± 1.4	37.0 ± 3.2
E_f (GPa)	0.73	1.8 ± 0.3	3.4 ± 0.3
Tensile properties			
σ_{max} (MPa)	3.4 ± 0.4	9.8 ± 1.2	12.5 ± 2.7
E_y (GPa)	0.23 ± 0.02	2.0 ± 0.1	2.0 ± 0.1
Remarks	Without backflow stop valve	Teflon® mould cavity	Teflon® mould cavity

consequently water evaporation. Materials obtained from injection moulding of the SFOC were fragile and needed to be handled with care (Table 7.6).

Thermomechanical and chemical treatments improve the flow behaviour of the SFOC (Figure 7.13) and hence, the amount of water necessary for forming decreases. Moulded materials from the ESTOC are denser and more resistant than those obtained from SFOC and ESFOC (Table 7.6). In addition, they could be formed in normal conditions used for the injection of thermoplastic materials, with the backflow stop valve.

The mechanical characteristics of SFOC-based materials, tensile and flexural stress at break values, respectively 12.5 MPa and 37 MPa, are slightly lower than those of commercial starch-based composite materials.[62] These materials are brittle when no external plasticiser is used, as in the case for all agromaterials.[50]

7.6 Durability of SFOC-Based Materials

The main advantage of protein-based materials, which is related to protein aggregation, is their high water resistance. While carbohydrate-based materials are highly moisture sensitive, moulded samples from ESTOC do not disintegrate when immersed in water at 25 °C and water absorption reaches a maximum of absorption lower than 60% in 24 h, while the samples injected from ESFOC cannot be handled after the same time (Figure 7.14). The protein matrix seems to have a thermoset-like behaviour rather than a thermoplastic behaviour. Heat- and shear-induced aggregation of proteins involving hydrophobic interactions, while the disulfide bridges generate a three-dimensional network.[26,63] This phenomenon is also responsible for the low proportion of water-soluble material in sunflower protein-based films obtained by thermomoulding.[37]

This essential property can be used in injection-moulding only because lignocellulosic fibres contained in the sunflower oil cake prevent the complete crosslinking of the proteins in the injection screw, as occurs during the forming of sunflower protein isolate when a temperature higher than its denaturation temperature is used.[64,65]

This specific property can be improved with a thermal treatment, which is known to reduce the hydrophilicity of wood or proteic films. In the former case, the treatment is based on the crosslinking of hemicelluloses and lignin compounds at high temperature and under inert atmosphere (200 °C, under nitrogen flow).[66] Above the glass-transition temperature of the amorphous

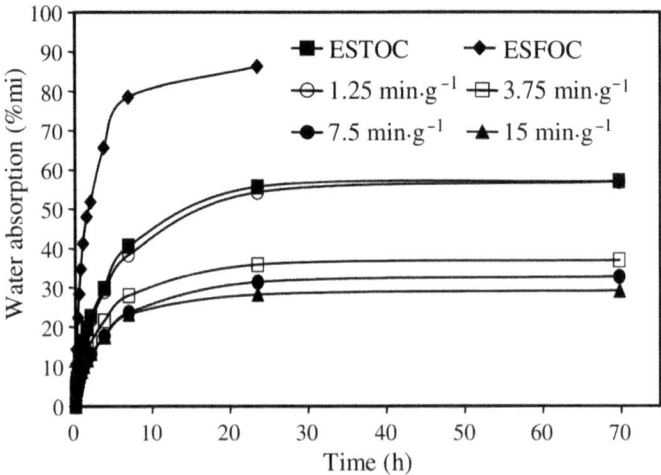

Figure 7.14 Mass gain in water at 25 °C of ESFOC and thermally treated ESTOC samples.[57]

biopolymers, this crosslinking reaction is allowed by the mobility of the biopolymer chains and the reactivity of phenolic compounds. The main consequence is the transformation of hemicelluloses, the most hydrophilic polymers, into a network with a more or less hydrophobic character. The amount of crosslinking is low, but the improvement of the physical properties of the wooden material is significant.[67]

In the case of protein films,[58,68] the treatment is carried out at lower temperatures (95–125 °C) than those used in the present study (200 °C), allowing the denaturation and the coagulation of proteins and the formation of new interactions (hydrophobic and cysteine bridges) in the loose network of films obtained by casting.

When applied to the ESTOC, the thermal treatment should have an influence only on fibres, because the protein network is well established. However, when phenolic compounds interact with proteins, they may form new bonds at higher temperatures, even in an inert atmosphere.

The first consequence of the thermal treatment is a loss of mass, which reaches 6.4% of the initial mass after equilibration at 25 °C and 60% R.H., and is associated with a decrease of volume. Finally, the apparent density decreases by about 0.1 during the first minutes of the treatment (Table 7.7). At the same time, the evaporation of water at this temperature results in irreversible modification of the treated samples.

Overall, thermal treatment leads to a decrease of the mechanical resistance of the specimens; the stress at break in bending and in tension decreases by 30% (Table 7.7), with an increase of their rigidity in the axis perpendicular to fibres, while the bending modulus increases during the treatment. The same consequences were observed during the treatment of wood.[66] This is related to the replacement of water plasticising molecules by interactions between polymeric chains.

The mechanical resistance loss (Table 7.7) is stabilised for a cooking duration higher than 3 min/g. For shorter times, the treatment involves first the disappearance of a part of the equilibrium absorbed water: at 1.25 min/g, the apparent density after equilibrium at 25 °C and 60% R.H. has already fallen to 1.2 and the mechanical resistance decreased greatly (Table 7.5). On the other hand, the kinetics of water absorption was practically unmodified (Figure 7.14), indicating that the treatment duration was not sufficient to form new covalent

Table 7.7 Mechanical properties in tension (σ_t : stress at break, E_y : Young modulus) and bending (σ_f : stress at break, E_f : bending modulus) of thermally treated ESTOC samples.[57]

Time (min/g)	0	1.25	3.75	7.5	15
Density	1.34 ± 0.03	1.21 ± 0.04	1.19 ± 0.06	1.20 ± 0.05	1.16 ± 0.04
σ_f (MPa)	37 ± 3	34 ± 4	24 ± 4	27 ± 5	25 ± 2
E_f (GPa)	3.3 ± 0.3	4.4 ± 0.6	3.8 ± 0.6	4.4 ± 0.2	4.2 ± 0.3
Time (min/g)	0	0.55	1.67	3.33	6.67
σ_t (MPa)	12.5 ± 2.7	11.5 ± 2.8	8.7 ± 2.6	7.8 ± 1.1	7.4 ± 1.8
E_y (GPa)	2.0 ± 0.1	2.1 ± 0.3	1.7 ± 0.3	1.9 ± 0.3	1.9 ± 0.5

Figure 7.15   SFOC-based moulded pieces.

Figure 7.16   Picture of horticultural trial of tomato plant growth in peat (left) and
SFOC-based pots (right).

bonds, but only new secondary interactions between chains. Above 3.75 min/g,
the mechanical properties were not further affected while water absorption
decreased; the absorption plateau is then found at less than 40%.

This finding has been applied to a practical application: the manufacture
of horticulture transplanting pots made from sunflower oil cake (Figures 7.15
and 7.16).

7.7   Compounding with other Plasticisers and/or Biopolymers

Even though moulding of SFOC, with a thermal–mechanical–chemical
treatment (referring to twin-screw extrusion and disulfide-bond reduction) is
processable on its own by injection-moulding; its "melt" viscosity is high and
the mechanical properties of formed materials are poor. This is the result of
high entanglement and aggregation of the protein clusters and the low ratio of
protein when compared to the whole SFOC.

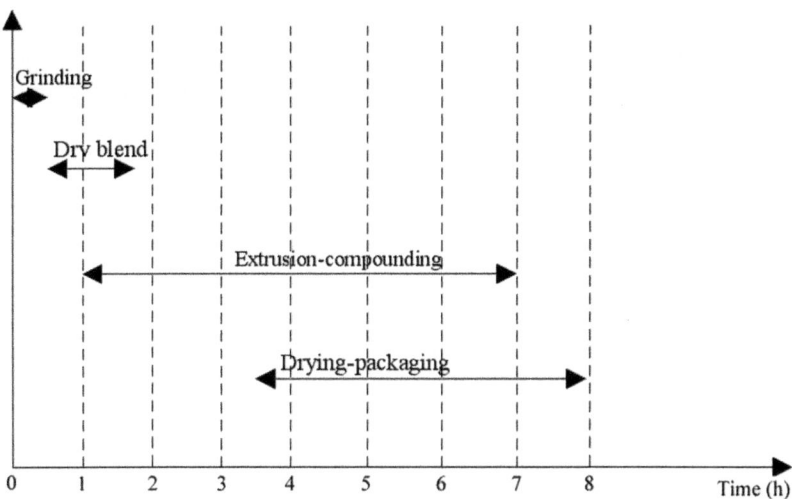

Figure 7.17 SFOC-based compounds production scheme.

Table 7.8 Flowing properties of SFOC-based compounds as measured in a dedicated spiral mould.

Agropellet	Composition	Spiral length (cm)
TE	SFOC	14.5 ± 0.6
TES	SFOC/sulfite	17.2 ± 1.7
TEG	SFOC/glycerol	16.0 ± 0.6
TEGS	SFOC/glycerol/sulfite	17.5 ± 0.1
TG	TEGS/gluten (100/15)	24.0 ± 0.6
TM	TEGS/corn meal (100/60)	46.4 ± 1.7
TL	TEGS/lignosulfonates (100/15)	36.7 ± 1.0

Operating conditions : 75%DM, $V_{inj} = 25$ mm s^{-1}, $P_{inj} = 900$ bar, $T_{inj} = 120\,°C$, $T_{mould} = 45\,°C$.

To enhance at the same time the processability and the material properties, which are clearly linked to their mechanical properties, and are directly related to density and also flowability (Table 7.6 and Figure 7.13), attempts have been made to add some other biopolymers.[70]

Among the large variety of available biopolymers, the addition of other crude agricultural products or byproducts, such as, corn flour, corn gluten and lignosulfonates was chosen.

The results of experiments made in a spiral mould, designed to monitor the flowing properties of compounds directly in injection-moulding, show clearly the positive effect of the coconstitutents (Table 7.8). Lignosulfonates behave as real rheological agents, and enhance greatly the filling of the mould and, thanks to its high content of starch, the corn meal increases the spiral length by a ratio of 3. In the chosen conditions for the comparison, the rheological properties of pure SFOC-derived compounds are poor but the effect is increased by the low injection pressure.

Table 7.9 Mechanical properties of moulded specimens according to compound composition. (G, L, M refer, respectively, to gluten, lignosulfonates and corn meal and are added in respective amounts of 15, 15 and 60 for 100 parts of ESFOC).

Compound	Tensile test			Bending test		
	E_Y (MPa)	σ_R (MPa)	ε_R (%)	E_f (MPa)	σ_R (MPa)	ε_R (%)
TE	1848 ± 124	8.49 ± 1.48	0.67 ± 0.03	1135 ± 85	11.5 ± 1.0	0.71 ± 0.05
TEG	2009 ± 112	8.69 ± 2.31	0.88 ± 0.29	1095 ± 66	10.5 ± 1.1	0.68 ± 0.04
TL	1523 ± 78	9.03 ± 1.15	0.76 ± 0.16	978 ± 62	10.7 ± 1.2	0.81 ± 0.06
TM	1584 ± 100	15.46 ± 2.16	1.01 ± 0.14	1482 ± 67	22.6 ± 0.6	1.00 ± 0.03
TGL	1644 ± 123	9.15 ± 0.47	0.88 ± 0.09	1044 ± 52	12.3 ± 1.0	0.84 ± 0.07
TGM	1965 ± 276	16.36 ± 1.35	1.06 ± 0.22	1379 ± 51	18.8 ± 0.9	0.91 ± 0.05
TLM	1507 ± 43	12.00 ± 1.23	1.04 ± 0.12	1199 ± 49	16.9 ± 1.4	0.95 ± 0.05
TGLM	1428 ± 78	11.80 ± 1.19	1.14 ± 0.21	1124 ± 35	17.7 ± 1.3	1.09 ± 0.08

Considering the mechanical properties of moulded pieces, the only enhancing bioproduct is the corn meal. Indeed, starch is well known to have a real thermoplastic behaviour[69] and to give good materials when compared to other biopolymers.[50] The addition of 60% of corn meal almost doubles the ultimate tensile strength (Table 7.9). Lignosulfonates have an interesting lubricating effect but do not affect mechanical properties. Corn gluten does not have any effect at all, indicating a behaviour towards thermomechanical processing similar to SFOC.

But in all cases, even if some these bioproducts improve some properties of SFOC-based materials, all the materials from these compounds failed the soaking test: they all lose their integrity in water. The crucial thermoset-like behaviour through disulfide bonds is hindered by the addition of copolymers.

7.8   Application Examples

The panel of applications is wide and different types of pieces have already been obtained (Figure 7.15).

As SFOC-based materials are, to a certain extent, water resistant, applications for horticulture are preferred. Manufacture of transplanting pots (as seen on the right of Figure 7.15) is presently being developed by a French company (AB7 Industries) with very good results. The pots withstand the 4-week germination period during which they are almost soaked in water. Then the whole, pot and plant, is set to ground without any transplantation stress nor any kind of toxicity. On the contrary, it was observed that SFOC degradation liberates available nitrogen and favors the plant growth (Figure 7.16).

7.9   Industrial Perspective

7.9.1   Compounding Cost Estimation

In a recent PhD thesis,[70] the cost estimation of the whole processing has been established for the production of flower pots from SFOC.

Figure 7.18   SFOC pellets.

This balance has been done using a compounding scheme based on a 8-h labour day for one operator (Figure 7.17) using the processing facilities of AGROMAT (French pilot platform dedicated to agromaterials).

- Grinding of SFOC pellets (Figure 7.18) is necessary to ensure a regular feeding of the twin-screw extruder. The ELECTRA hammer-mill used was fitted with a 2-mm grid and is designed for an 800 kg/h capacity. Electrical consumption is estimated at 5.5 kWh/t by the manufacturer. The grinding of 380 kg of SFOC for the production trial was performed in 30 min, at a consumption of 2.30 kWh.
- If a dry blend is needed, for example, for the addition of another solid bioproduct, a double-helix horizontal blender ELECTRA MH 400 was used. 100 kg batches were realised in 15 min at an electrical power of 3 kW.
- Twin-screw extrusion was performed on a CLEXTRAL EVOLUM HT 53. 7 h were necessary for the production of 380 kg TEGS (SFOC input: 55 kg/h), with a mechanical energy of 167 kW/kg. For this trial, the overall capacity at this stage was limited by the solid and liquid inputs. The production could be increased by a 2 to 3 ratio as the screw speed was only 300 rpm out of a 800 rpm maximum.
- The extruded compounds moisture content was around 25% with drying performed on a continuous belt dryer CLEXTRAL EVOLUM 600.

Table 7.10   Costs of raw materials, energy and labor
used for the cost calculation.

Sunflower oil cake (SFOC)	0.25 €/kg
Water	0.00309 €/kg
Glycerol	0.4 €/kg
Sodium sulfite	1.055 €/kg
Energy	0.1057 €/kW
Labour	10.50 €/h

Table 7.11   Operating expenses for the production of
SFOC-based compounds.

	TEGS
Raw materials (€)	116
Utility (€)	16.5
Labor time (€)	84
Operating costs (€)	214
Production (kg)	380
Final cost (€/kg)	0.56

Operating conditions for the 3.2 m^2 drying surface were a temperature of 50 °C, for a residence time of 20 min, until the moisture content reached 4–6%. Energy consumption was estimated at 40 kW for the drying of 460 kg of wet compound. This stage is mandatory for the preservation of the compound. Direct injection moulding of the wet compounds could be interesting and should involve a reduction of the energy input but, the whole process would then need regauging to adapt the compounding capacity to the moulding capacity.

The final balance calculated from raw materials, energy and labour time gathered in Table 7.10, the production of SFOC-based compounds TEGS (containing 8% of glycerol, 2% of sodium sulfite), without the equipment depreciation, is estimated at 0.56 €/kg (Table 7.11).

The study gives only an approximate range for the SFOC-based compounds but it is obvious that this kind of biocomposite can be very interesting when compared to the cheapest thermoplastics (i.e. PE = 1.3 €/kg, PP = 1.2 €/kg, PS = 1.2 €/kg).

7.9.2   Injection Moulding Cost Estimation

Injection moulding costs have additionally been evaluated for the production of the transplanting pots. Their production was performed on a NEGRI BOSSI VE 160-720 press with the operating conditions summarised in Table 7.13. With a 30-s cycle time, injection-moulding rate of production can reach

Table 7.12 Cost estimate for the moulding of one planting out pot from SFOC.

	TEGS
Raw materials (€/kg):	0.56
Pot weight (g)	80
Raw material cost (€/pot)	0.045
Energy cost:	
*kW/pot	0.01
*€/pot	0.1
Labor cost (€/pot)	0.021
TOTAL (€/pot)	**0.076**

Table 7.13 Operation conditions for the production of flower pots from SFOC.

Parameter	Value
Shot build:	
✓ length (mm)	62
✓ speed (rpm)	100
✓ counter pressure (bars)	15
Injection:	
✓ pressure (bars)	1000
✓ speed (mm.s^{-1})	50
Holding:	
✓ pressure (bars)	50
✓ duration (s)	3
✓ commutation length (mm)	10
Temperatures	
✓ mold (°C)	20
✓ nose/zone 3/zone 2/zone 1 (°C)	120/110/90/75
Cooling time (s)	20

120 pots/h. The estimate of the total manufacture cost for one pot is then: 0.076 € (Table 7.12).

This calculation has only been made based on semi-industrial trials, but gives a good indication of the production rate and the price range that can be reached for such pots.

7.9.3 Environmental Benefits of Natural Biocomposites

The main environmental benefit of SFOC-based materials is their biodegradability. This was clearly demonstrated previously in the horticultural trials involving the growth of tomato plants, which also verified the nonecotoxicity of the degradation products (for tomatoes at least).

Table 7.14 Mineralisation of protein injected specimens (ISi) and thermomoulded films (TFi). IC: Inorganic carbon dissolved; Inh.ISi: Inhibitor of sample ISi; Inh.TFi: Inhibitor of Sample Tfi.

Time (days)	IS1	IS2	Inh.IS1	TF1	TF2	Inh.TFi	Cellulose
27	71.8%	64.7%	78.5%	65.2%	55.5%	75.5%	67.5%
41	74.8%	67.6%	81.2%	68.3%	58.6%	78.0%	70.4%
$CO_2 + IC$	75.5%	68.2%	81.6%	70.0%	59.0%	78.5%	71.1%
Average		**71.8%**	–		**64.5%**	–	–
SD		5.2%	–		7.7%	–	–

Biodegradability of sunflower protein-based materials has also been tested in standardised liquid conditions EN 13432 (Table 7.14). Injected and thermo-moulded samples are ranged, respectively, as easily biodegradable and inter-mediately biodegradable, with final mineralisation rates of 71.8% and 64.5% with no inhibition effect of these materials on the microbial population.

Another aspect is the life-cycle assessment of this type of composite. Unfortunately, the LCA focus has been made on bioplastics such as PLA,[70] and not on natural biocomposites. However, some insight has been presented in a recent evaluation by the VEGEPLAST® company. Carbon balance (Bilan Carbone® from the French environmental agency ADEME) of this cereal-based material showed that the use of regional crops and basic thermomech-anical processes gives a low CO_2 emission equivalent to 0.76 tons per ton, lower than all conventional plastics. SFOC-based materials, being transformed similarly, should then present almost the same benefit balance.

Conclusion

Oilseed proteins do possess interesting physical and chemical properties for nonfood applications especially in the field of biobased materials.

Resources are huge and a complete nonfood culture could be more intensively developed in fallow lands, as oil is a promising biochemical for fuel and lubricant applications.

Unfortunately, if soy proteins are clearly identified and used industrially (mainly for food processes), industrial extraction of proteins from other crops (*i.e.* sunflower, rapeseed, linseed) is still in development. This is the main brake on the industrial use of oil cakes, as a better understanding of proteins real potential is needed. Their complicated behaviour under shear and heat, involving a broad range of secondary interactions and covalent crosslinking,[17] is not yet really controlled, yet shows great promise. Protein-based adhesive and materials have already showed some particularly interesting properties in composites and particle boards.[46–48]

Thermomechanical transformation of raw oil cakes (as shown with the example of sunflower oil cake) is a promising process due to:

- the occurrence of lignocellulosic fibres avoids the complete coagulation of proteins and facilitates processes like extrusion or injection moulding;

- the mechanical properties of oil cake-based materials are lower than for similar starch-based composites but they possess a natural resistance to moisture that should broaden the field of applications, especially for horticulture;
- the economic and environmental assessments of oil-cake-based materials are definitely positive.

The overall balance for the biomaterials production from oil cakes is already good. To make it even better some specific points should be studied:

- rapeseed and linseed should be tested and a comparative study of oil seed proteins from a material point view could be very interesting to evaluate the real potentials and the influence of some specific factors: polyphenolic compounds, amino acid composition, specific constituent (*e.g.* mucilage, starch, *etc.*),
- the rheological behaviour of oilseed protein in the "molten" state is not controlled (crosslinking, kinetic factor, *etc.*) while it could really create a breakthrough in nonfood applications,
- improvement of oil cake-based materials should be investigated, the protein enrichment of the oil extraction byproduct could be an interesting way to do it.

References

1. De la production à la consommation. Statistiques des oléagineux et protéagineux 2011–2012. Prolea, Paris, 2012.
2. T. M. Lammens, M. C. R. Franssen, E. L. Scott and J. P. M. Sanders, *Biomass Bioenergy*, 2012, **44**, 168.
3. P. R. Salgado, S. E. Molina Ortiz, S. Petruccelli and A. N. Mauri, *J. Am. Oil Chem. Soc.*, 2011, **88**, 351.
4. M. Aider and C. Barbana, *Trends Food Sci. Technol.*, 2011, **22**, 21.
5. P. R. Salgado, M. Elvira Lopez-Caballero, M. Carmen Gomez-Guillen, A. N. Mauri and M. Pilar Montero, *Food Hydrocolloids*, 2012, **29**, 374.
6. A. Nesterenko, I. Alric, F. Silvestre and V. Durrieu, *Ind. Crops Prod.*, 2013, **42**, 469.
7. P. R. Salgado, S. R. Drago, S. E. Molina Ortiz, S. Petruccelli, O. Andrich, R. J. Gonzalez and A. N. Mauri, *LWT-Food Sci. Technol.*, 2012, **45**, 65.
8. Y. X. Niu, W. Li, J. Zhu, Q. Huang, M. Jiang and F. Huang, *Int. J. Food Eng.*, 2012, **8**.
9. G. H. Brother and L. L. McKinney, *Ind. Eng. Chem.*, 1939, **31**, 84.
10. H. Singh, *Trends Food Sci. Technol.*, 1991, **2**, 196.
11. R. Kumar, V. Choudhary, S. Mishra, I. K. Varma and B. Mattiason, *Ind. Crops Prod.*, 2002, **16**, 155.
12. S. Damodaran, in *Food Proteins and Their Applications*, Marcel Dekker, NY-Basel, 1996, pp. 1–24.

13. C. K. Mathews, K. E. Van Holde, and K. G. Ahern, *Biochemistry 3rd edn*, Benjamin-Cummings, 1999.
14. S. Damodaran, in *Food Chemistry*, Marcel Dekker Inc., NY-Basel, 1996, vol. third, pp. 321–431.
15. S. Gonzalez-Perez, J. M. Vereijken, K. B. Merck, G. A. van Koningsveld, H. Gruppen and A. G. J. Voragen, *J. Agric. Food Chem.*, 2004, **52**, 6770.
16. A. Rouilly, O. Orliac, F. Silvestre and L. Rigal, *Thermochim. Acta*, 2003, **398**, 195.
17. C. Tang, S. Choi and C. Ma, *Int. J. Biol. Macromol.*, 2007, **40**, 96.
18. L. A. De Graaf, *J. Biotechnol.*, 2000, **79**, 299.
19. L. Slade and H. Levine, in *Water Relationships in Food*, Plenum Press, NY and London, 1991, pp. 29–101.
20. A. Morales and J. L. Kokini, *Biotechnol. Prog.*, 1997, **13**, 624.
21. C. Kealley, M. Rout, M. Dezfouli, E. Strounina, A. Whittaker, I. Appelqvist, P. Lillford, E. Gilbert and M. Gidley, *Biomacromolecules*, 2008, **9**, 2937.
22. J. Zhang, P. Mungara and J. Jane, *Polymer*, 2001, **42**, 2569.
23. A. Rouilly, O. Orliac, F. Silvestre and L. Rigal, *Polymer*, 2001, **42**, 10111.
24. J. A. G. Arêas, *Crit. Rev. Food Sci. Nutr.*, 1992, **32**, 365.
25. C. J. R. Verbeek and L. E. van den Berg, *Macromol. Mater. Eng.*, 2010, **295**, 10.
26. A. Redl, S. Guilbert and M. H. Morel, *J. Cereal Sci.*, 2003, **38**, 105.
27. D. Liu, H. Tian and L. Zhang, *J. Appl. Polym. Sci.*, 2007, **106**, 1307.
28. O. Orliac, F. Silvestre, A. Rouilly and L. Rigal, *Ind. Eng. Chem. Res.*, 2003, **42**, 1674.
29. T. A. Egorov, T. I. Odintsova, A. K. Musolyamov, R. Fido, A. S. Tatham and P. R. Shewry, *FEBS Lett.*, 1996, **396**, 285.
30. M. F. Marcone, Y. Kakuda and R. Y. Yada, *Food Chem.*, 1998, **63**, 265.
31. E. C. Y. Li-Chan and C.-Y. Ma, *Food Chem.*, 2002, **77**, 495.
32. U. Kalapathy, N. S. Hettiarachchy and K. C. Rhee, *J. Am. Oil Chem. Soc.*, 1997, **74**, 195.
33. M. H. Morel, J. Bonicel, V. Micard and S. Guilbert, *J. Agric. Food Chem.*, 2000, **48**, 186.
34. I. Paetau, C.-Z. Chen and J.-L. Jane, *Ind. Eng. Chem. Res.*, 1994, 1821.
35. S. Wang, S. Zhang and J.-L. Jane, *J. Macromol. Sci. - Pure Appl. Chem.*, 1996, **AA33**, 557.
36. C. H. Schilling, T. Babcock, S. Wang and J. Jane, *J. Mater. Res.*, 1995, **10**, 2197.
37. O. Orliac, A. Rouilly, F. Silvestre and L. Rigal, *Polymer*, 2002, **43**, 5417.
38. O. Orliac, A. Rouilly, F. Silvestre and L. Rigal, *Ind. Crops Prod.*, 2003, **18**, 91.
39. H. C. Huang, T. C. Chang and J. Jane, *J. Am. Oil Chem. Soc.*, 1999, **76**, 1101.
40. P. Nayak, A. Sasmal, P. Nayak, S. Sahoo, J. Mishra, S. Kang, J. Lee and Y. Chang, *Polym.-Plast. Technol. Eng.*, 2008, **47**, 600.

41. J. Zhang, L. Jiang and L. Zhu, *Biomacromolecules*, 2006, **7**, 1551.
42. P. Mungara, J. Zhang, S. Zhang, and J.-L. Jane, *Protein-Based Films and Coatings*, A. Gennadios, CRC Press, Boca Raton, London & New York, 2002, 621–638.
43. A. K. Mohanty, P. Tummala, W. Liu, M. Misra, P. V. Mulukutla and L. T. Drzal, *J. Polym. Environ.*, 2005, **13**, 279.
44. C. M. Vaz, P. F. N. M. Van Doeveren, R. L. Reis and A. M. Cunha, *Biomacromolecules*, 2003, **4**, 1520.
45. A. Rouilly, A. Meriaux, C. Geneau, F. Silvestre and L. Rigal, *Polym. Eng. Sci.*, 2006, **46**, 1635.
46. Y. Fahmy, N. A. El-Wakil, A. A. El-Gendy, R. E. Abou-Zeid and M. A. Youssef, *Int. J. Biol. Macromol.*, 2010, **47**, 82.
47. S. Khosravi, F. Khabbaz, P. Nordqvist and M. Johansson, *Ind. Crops Prod.*, 2010, **32**, 275.
48. S. Khosravi, P. Nordqvist, F. Khabbaz and M. Johansson, *Ind. Crops Prod.*, 2011, **34**, 1509.
49. G. Chabanon, I. Chevalot, X. Framboisier, S. Chenu and I. Marc, *Process Biochem.*, 2007, **42**, 1419.
50. A. Rouilly and L. Rigal, *J. Macromol. Sci.-Polym. Rev.*, 2002, **C42**, 441.
51. A. Rouilly, O. Orliac, F. Silvestre, and L. Rigal, in *12th European Conference and Technology Exhibition on Biomass for Energy, Industry and Climate Protection*, Amsterdam, Netherland, 2002.
52. C. Geneau-Sbartai, J. Leyris, F. Silvestre and L. Rigal, *J. Agric. Food Chem.*, 2008, **56**, 11198.
53. J. Raymond, N. Rakariyatham and J. Azanza, *Sciences des aliments*, 1985, **4**, 95.
54. M. C. S. Sastry and M. S. N. Rao, *J. Agric. Food Chem.*, 1991, **39**, 63.
55. S. Gonzalez-Perez and J. Vereijken, *J. Sci. Food Agric.*, 2007, **87**, 2173.
56. S. N'Diaye and L. Rigal, *Bioresour. Technol.*, 2000, **75**, 13.
57. A. Rouilly, O. Orliac, F. Silvestre and L. Rigal, *Bioresour. Technol.*, 2006, **97**, 553.
58. V. Micard, R. Belamri, M. H. Morel and S. Guilbert, *J. Agric. Food Chem.*, 2000, **48**, 2948.
59. J. L. Willett, B. K. Jasberg and C. L. Swanson, *Polym. Eng. Sci.*, 1995, **35**, 202.
60. R. Parker, A.-L. Ollett, and A. C. Smith, in *COST 91 bis*, Elsevier, Gottenburg, 1989.
61. V.-M. Archodulaki and N. Mundliger, *Kunstoffe*, 1997, **87**, 634.
62. P. Krajewsky and H. Patzschke, *Kunstoffberater*, 1999, **44**, 35.
63. G.-T. Meng, K.-M. Ching and C.-Y. Ma, *Food Chem.*, 2002, **79**, 93.
64. Y. Wang, A. M. Rakotonirainy and G. W. Padua, *Starch/Staerke*, 2003, **55**, 25.
65. O. Orliac, F. Silvestre, A. Rouilly and L. Rigal, *Ind. Eng. Chem. Res.*, 2003, **42**, 1674.
66. R. Guyonnet, 1998, French Patent FR 2 604 942.

67. B. F. Tjeerdsma, M. Boonsta, A. Pizzi and H. Militz, *Holz Roh- Werkst.*, 1998, **56**, 149.
68. A. Gennadios, V. M. Ghorpade, C. L. Weller and M. A. Hanna, *Trans. ASAE*, 1996, **39**, 575.
69. R. F. T. Stepto, *Macromol. Symp.*, 2000, **152**, 73.
70. J. Humbert, PhD Thesis, INP Toulouse, 2008.

Subject Index

References to figures are given in *italic* type. References to tables are given in **bold** type.